JOHN DEWEY
AND
ENVIRONMENTAL
PHILOSOPHY

SUNY series in Environmental Philosophy and Ethics
J. Baird Callicott and John van Buren, editors

JOHN DEWEY AND ENVIRONMENTAL PHILOSOPHY

Hugh P. McDonald

STATE UNIVERSITY OF NEW YORK PRESS

Published by
State University of New York Press, Albany

© 2004 State University of New York

All rights reserved

Printed in the United States of America

No part of this book may be used or reproduced in any manner whatsoever without written permission. No part of this book may be stored in a retrieval system or transmitted in any form or by any means including electronic, electrostatic, magnetic tape, mechanical, photocopying, recording, or otherwise without the prior permission in writing of the publisher.

For information, address State University of New York Press,
90 State Street, Suite 700, Albany, NY 12207

Production by Kelli Williams
Marketing by Michael Campochiaro

Library of Congress Cataloging-in-Publication Data

McDonald, H. P. (Hugh P.)
 John Dewey and environmental philosophy / Hugh P. McDonald.
  p. cm. — (SUNY series in environmental philosophy and ethics)
 Includes bibliographical references and index.
 ISBN 0-7914-5873-3 (alk. paper) — ISBN 0-7914-5874-1 (pbk. alk. paper)
  1. Dewey, John, 1859–1952. 2. Pragmatism. 3. Environmental sciences—Philosophy. 4. Environmental ethics. I. Title. II. Series.

B945.D44M34 2003
179'.1'092—dc22

2003060825

10 9 8 7 6 5 4 3 2 1

*This book is dedicated to the late
Professor Reuben Abel,
teacher, scholar, philosopher*

CONTENTS

ACKNOWLEDGMENTS		xi
PREFACE		xiii
ONE	Environmental Ethics and Intrinsic Value	1

PROLOGUE TO CHAPTER TWO — 57
The Setting of the Problem of Pragmatism and the Environment:
The Critique of Pragmatism as an Environmental Ethics in
Taylor, Bowers, Katz, and Weston

TWO	Dewey's Naturalism	67
THREE	Dewey's Instrumentalism	91
FOUR	Dewey's (Moral) Holism	109
FIVE	Dewey's Ethics as a Basis for Environmental Issues	123
EPILOGUE	Pragmatism and Environmental Ethics	143
NOTES		159
BIBLIOGRAPHY		207
INDEX		215

Human nature exists and operates in an environment . . . as a plant is in the sunlight and soil. It is of them, continuous with their energies, dependent upon their support, capable of increase only as it utilizes them and as it gradually rebuilds from their crude indifference an environment genially civilized.

—Dewey

ACKNOWLEDGMENTS

My thanks to Professor Richard Bernstein, who patiently read the initial drafts of this work and provided many helpful comments and criticisms. In particular, his suggestion that pragmatism might provide an alternative approach to environmental ethics, apart from intrinsic value, proved fruitful and changed the direction of the book.

Chapter two, Dewey's Naturalism, has appeared in *Environmental Ethics* in a slightly altered form.

I also wish to acknowledge the influence of J. Baird Callicott on my thinking, despite many critical comments in chapter two.

Reprinted from *In Defense of the Land Ethic: Essays in Environmental Philosophy*, by J. Baird Callicott, by permission of the State University of New York Press, © 1989, State University of New York. All rights reserved. Portions throughout.

From excerpts in *Environmental Ethics, Duties to and Values in the Natural World*, by Holmes Rolston III. Reprinted by permission of Temple University Press. © 1988 by Temple University. All Rights Reserved. Portions throughout.

From excerpts in *Experience and Nature*, by John Dewey. Grateful acknowledgment is made to Open Court Publishing Company, a division of Carus Publishing Company. © 1925, 1929 by Open Court Publishing Company.

PREFACE

Environmental ethics has recently developed as a separate field within philosophy and has a large and growing literature of its own. This field involves ethical studies of the rights of other species, of ecosystems, and of the environment of all living things as a whole. Much of the internal debate within the field has concerned issues of value and moral considerability. It is argued that if species other than humans, ecosystems, and the biosphere—the entire planetary zone of life—have value on their own, then they are entitled to moral consideration from humans, either some form of rights, or a duty of humans to the environment. This may take the minimal form of letting the bearer of rights alone or a larger, more activist role in managing the environment with an eye to its benefit.

The central issue in this debate, which has attracted perhaps the most attention in the journals, is the issue of the value of the environment, other species, and ecosystems. One of the interesting things about this ongoing discussion has been the almost exclusive concentration on issues of value in the literature of environmental ethics. This has revived value studies, which had shown recent signs of fading,[1] in a vigorous way. The debate has been framed in terms of "anthropocentric" as opposed to "non-anthropocentric" values.[2] "Anthropocentric" values are values that are exclusive to humans in some respect, whether involving a uniquely human capacity such as some would claim reason to be; or the purported unique moral agency of humans, that humans alone are rights bearers, agents of duty, or the like. Non-anthropocentrists locate value in a larger set of members or a set with larger scope. This may include higher animals, all living things, ecosystems, species, and the biosphere. The value structure that has framed the debate has not so much challenged traditional divisions of values, namely, instrumental and intrinsic values, as the exclusivity of human value and human valuing. Non-anthropocentrists have argued that anthropocentric value theories, as their etymology would imply, are oriented exclusively toward humans. They are arbitrary and "speciesist" in their view of value as based solely in human being and as excluding all other species and the environment itself from moral consideration. Other species and the environment are treated exclusively as instruments for human use, and not given consideration in their own right as worthy of preservation, survival, and autonomy. Because, it is charged, traditional Western ethics is almost universally anthropocentric, a new, environmental ethic is needed, which will challenge the primacy or even exclusivity of humans to ethical consideration.[3] It is not enough for some of the radicals in this movement to merely rework traditional ethics with an eye to the environment. Environmental ethics is considered by some to be a revolution in ethics, which will completely change the human-centered orientation of morals to include all species, ecosystems, or the larger biosphere.

The critique of Western philosophy as a whole has been extended by individual authors to criticisms of particular schools of philosophy in the West, for example, utilitarianism and other prominent ethical theories. Recently, there have been criticisms of pragmatism as a whole, as well as more specific criticisms of James and Dewey, two of its more accessible and prominent members.[4] Pragmatism has been criticized as being inherently anthropocentric and thus implicitly anti-environmental in the strong sense that environmentalism cannot be anthropocentric. These criticisms of pragmatism have as a premise that all the various authors who have been identified as pragmatists have some common elements that make them pragmatists and that these are implicitly hostile or indifferent to environmentalism or environmental ethics. More specifically, the instrumental theory of values in Dewey is regarded as an implicit critique of the very possibility of assigning or locating intrinsic value in any of the candidates for such a larger set or locus of intrinsic value, for example, higher animals or ecosystems. Thus pragmatism, it is alleged, would undermine the very possibility of an environmental ethic by arguing against its central and ultimate foundation, intrinsic value, if it were to be used as the basis for environmental ethics.

My main argument will be that the pragmatism of Dewey can be defended against charges of anti-environmentalism and anthropocentricity.[5] Dewey was picked to represent pragmatism not only because of his highly developed and intricate theory of values and his naturalistic approach but because his naturalism already is environmentally minded. In order to investigate the critique of pragmatism by certain authors from the point of view of environmental ethics, I will also include an analysis of these criticisms as a prologue to chapter two.

In chapter one, I will analyze several representative philosophies that argue for the intrinsic value of nature: those of J. Baird Callicott, Tom Regan, and Holmes Rolston III. These are three of the most prominent environmental philosophers, who have published the most thoroughgoing and articulate theories of intrinsic value in environmental philosophy. This chapter will introduce and critically examine the concept of the intrinsic value of nonhuman life and the ecology.[6] I will also consider a prominent critic of non-anthropocentric intrinsic value theories from within the environmental literature, Bryan Norton.

The next four chapters will present a detailed analysis of the (practical) philosophy of John Dewey. However, I will also include a brief examination of some elements of his naturalism, especially in chapter two. The second chapter is in a sense central to the book as it directly refutes the notion that pragmatism could not serve as an environmental ethic. I argue that Dewey brings human nature back into nature as a whole, ending the isolation and detachment of the subject. Values are treated along with natural processes in a way that provides an alternative model for environmental ethics, one that is more closely in accord with the formal requirements of such an ethic. Dewey's naturalism could be used, I will argue, for such an alternative model. Thus the criticism of pragmatism as a whole as undermining the central foundation of environmental ethics will be challenged.

In chapter three, a prominent and thoroughgoing theory of value by a pragmatist, John Dewey, will be analyzed. Dewey's criticisms of the whole notion of intrinsic value will be closely examined, to determine if he undermines it, as a prominent environmental ethicist has implied. I will argue that Dewey actually refines intrinsic value along naturalistic lines. This chapter will also involve correcting or setting the record straight both about Dewey and about pragmatism.

In chapter four, Dewey's holism in method and its use as a framework and justification for ethics is examined. The place of intrinsic value in a holistic model of ethics is examined in some detail. Thus the difference between Dewey's approach to value and that of other consequentialist theories will be analyzed. Although Dewey's position is not as radical as that of some of the recent authors in environmental ethics, his position is close to that of several of the non-anthropocentrists and less anthropocentric than others. Indeed, in many important particulars, it is quite similar to one of the more radical figures, H. Rolston. These points of similarity are close in spirit, it will be argued, to that of the whole movement of environmental philosophy.

In chapter five, Dewey's theory will be applied to issues in environmental philosophy. While chapter four argues that Dewey was a holist in ethics, chapter five applies this model to actual issues. His position on values and his overall ethical naturalism will be used to determine what he might have said about recent issues in environmental ethics. This will draw on scattered remarks Dewey made on issues such as the human treatment of animals as well as the extension of his principles to other issues in environmental philosophy. Thus the chapter speculates on just how a Deweyan approach will help in dealing with environmental issues.

In the final epilogue, an alternative to the present debate in environmental ethics from the point of view of pragmatism will be presented. This section will not contain the sustained analysis of the earlier chapters: it is only meant to be suggestive. The argument of this section will be that certain approaches taken from pragmatism as a whole would provide an alternative model for environmental ethics. Indeed, pragmatism as a whole may provide both a more defensible theory of intrinsic or "inherent" value than other ethical schools, and, more generally, an alternative model of what environmental philosophy could be. The subjectivity of values based on certain specific psychological capacities or subjective states of humans has been undermined by Peirce's notion of community and Mead's analysis of the social influences on the formation of individual minds. Peirce and C. I. Lewis not only uphold a notion of intrinsic value, but Lewis also provides a better foundation for a relational theory than such naturalists as Callicott. The holism of most of the pragmatists constitutes a challenge to value and ethics centered in the individual, and a possible ground for more holistic theories of value, which, some have argued,[7] is more suitable to an environmental, as opposed to a humane, ethic. As the world is not devalued in the subject-object dichotomy for the pragmatists, the possibility of valuing the nonhuman is less difficult and more consistent with its theories of value. Although there are points of difference between them, there are central points of similarity, which are very suited to environmental ethics. They all offer a life-centered theory of value, which is not subjective; a holistic method; and an alternative model of the relation of value and obligation.[8] The pragmatists provide an alternative, within consequentialism, to utilitarianism and other philosophies centered on the subject.

The metaphysics of the subject is the main source for the devaluation of nature. Pragmatism, without exception, has been a movement away from subjectivity. I will examine the criticisms of William James and use James' texts to argue that criticisms of James as a subjectivist involve misinterpretations. Still other pragmatic innovations might, taken as a whole and properly reworked, provide an alternative to the almost exclusive emphasis on intrinsic value in environmental ethics. Most of the pragmatists have argued against the strong separation of fact and value, which marks the division between the world of nature and man and thus the ground of strongly anthropocentric ethics. On a speculative level, Peirce's pan-psychic or "objective" idealism is close to the spiritual reading of nature in some "deep ecologists," and the "Gaia" hypothesis. Another point of departure might be the "transactionalism" of Dewey's late period and his collaboration with Arthur Bentley. Transactionalism is a refinement of the anti-subjective movement of pragmatism, since the "subject" is constituted as events in transaction. In this view the subject is joined to the environment even more closely, since the subject is not removed from the environment in a "hypostatization," but is a series of events within it. Finally, Schiller's foundational model of ethics, although "humanist," could, combined with the other approaches of pragmatism, liberate environmental ethics from ontological and metaphysical commitments foreign to it. The attempt to establish environmental ethics as a revolutionary change of outlook in ethics, which breaks with traditional Western philosophy, and presumably its metaphysical ethics, might use Schiller's framework to establish its autonomy.

I will argue not only that pragmatism as a whole avoids the subjectivity and anthropocentricity of modern Western philosophy and provides a carefully worked out theory of intrinsic value. With extension, it could also provide an alternative model for environmental ethics that would be more compatible with the requirements of such a philosophy.

One note on method. I am maintaining a coherentist interpretation of Dewey in this book. Except for his early, Hegelian period, I believe that the many works he wrote are internally coherent and consistent. Most of the works of the "early" period (at least the post-Hegelian part of it), the middle period, and the late period can be made consistent with one another.[9] I will argue for such a coherentist reading in some of the notes to chapters three and four in response to other readings of specific issues, for example, Dewey's consequentialism. However, there are two main problems with such a coherentist interpretation.

The first is the movement in Dewey's late period towards "transactionalism," which was mentioned earlier.[10] If Dewey was moving away from the major works of his life in this very late period, does this constitute evidence against any coherentist view? Is there a "break" between the major works of Dewey's mature period and the transactionalism of his last years? Commentators have differed. Some have argued that Dewey's collaboration with Bentley in the *Knowing and the Known* "dissolves" and thus threatens to undermine all of Dewey's earlier projects and works.[11] Others have seen more continuity, arguing that "the ideas presented are all to be found in Dewey's work."[12] Still others have argued that the work contains far more of Bentley than of Dewey and thus must be approached cautiously. Indeed, one commentator argued that Dewey was "reluctant to sign on as co-author, since most of the work was Bentley's."[13] In this last view, Dewey was only a provisional transactionalist.

I would maintain that, except in certain details, the transactional period is consistent with the rest of Dewey's work. However, I will not use *Knowing and the Known* as a source text in the bulk of what follows. There are several reasons for this omission. Firstly, there is no consensus on which part of the work is Bentley's and which Dewey's. This problem also occurs with other Deweyan collaborations, of course, namely his collaboration with Tufts on the *Ethics*. However, the chapters of the latter work were clearly assigned to one author or another. While this holds for certain chapters of *Knowing and the Known*, many others are, at least ostensibly, conjoint.[14] Since many of the chapters are Bentley's alone, and all but one of the others were written by

Bentley and then reworked by Dewey, it would be difficult to separate out what is clearly Dewey's from the work. Second, Dewey probably signed on because he thought the book consistent with what he did before for the most part. In other words I agree with Ratner that despite differences in vocabulary, the thrust of transactionalism as a nonsubjective process philosophy can be reconciled with *Experience and Nature* and other works of Dewey's mature years. Since the others are undoubtedly Dewey's, there is no problem of separation. Third, since the work was primarily Bentley's, even though Dewey signed on, it is not an unambiguously Deweyan text. While he approved and perhaps reworked much of Bentley's material, the bulk of the book is Bentley's. *Knowing and the Known* is consistent with Dewey's work, but that does not make it Dewey's work. I believe some of what William James wrote also is consistent with what Dewey wrote. However, I would not attribute a Jamesian doctrine to Dewey without textual support. In this book, I am reluctant to incorporate material that Dewey had reservations about and may not be his own. But I will return to this text in the Epilogue, since it may provide still another pragmatic framework from which an environmental ethic could be developed.

The second main challenge to a "coherentist" interpretation of Dewey's works is the revisionist reading of Dewey contained in the works of Richard Rorty, primarily his *Consequences of Pragmatism*.[15] Rorty quotes a letter from Dewey to Bentley in which the former regrets the use of the word "experience" in *Experience and Nature*, Dewey's main work on metaphysics. Rorty sees this as evidence of a movement away from a metaphysics by Dewey in his late years, since the repudiation of the metaphysics of experience could be taken as an abandonment of metaphysics. What Rorty does not point out is that this letter was written in the context of writing *Knowing and the Known*. Yet Rorty also eschews the transactionalism of the latter work.[16] In other words, Dewey was not moving toward the "therapeutic" approach to philosophy championed by Rorty, but toward a more "event" oriented metaphysics, transactionalism. However, I do not accept his idiosyncratic reading of Dewey, which argues that most of the major works of Dewey were a mistake and that Dewey can properly be interpreted from the (Epicurean-)Wittgensteinian perspective of philosophy as therapy.

Firstly, as I previously argued, I believe that all of Dewey's works after the early Hegelian period are consistent, including his transactionalism. Even if the latter were to represent a radically new approach, or departure, however, it is metaphysical. Thus Rorty's attempt to reinterpret Dewey as anti-metaphysical fails.[17] Rorty may not like Dewey's metaphysics, since it constitutes strong evidence against his therapeutic reading of Dewey, but the question is whether Rorty's is an adequate reading of Dewey. Dewey was a systematic philosopher, contra Rorty, and experience played a central role in Dewey's system. Where Rorty is correct, I think, is in pointing out Dewey's regret over using such a word as "experience." I will address this point in more detail in the Epilogue, but it is true that Dewey attempted to rework the word away from its subjective connotations. He realized near the end of his life that this was an impossible and fruitless task.

I know of no text of Dewey's in which he espouses a "therapeutic" point of view and, to the contrary, I believe he would have rejected this interpretation. Philosophy as therapy involves a contemplative view or perspective, not a practical one.[18] Dewey would have seen the movement to make philosophy therapeutic as a return to the contemplative ideal, and the divorce of theory from practice. Regardless of the metaphysical issues, this is not Dewey, early or late. Rorty is a methodological pragmatist—that is, he approaches issues with a pragmatic outlook, perspective, and edge. He is not, however, a doctrinal pragmatist and does not like systems, formality, and the whole enterprise of metaphysics, despite his own materialism. He has gone through the "linguistic turn" and thus does not really share the metaphysical perspective of the pragmatists, many of whom began from experience. Rather, it seems to me that Rorty has articulated a new form of pragmatism, allied to materialism.[19] While this shares Dewey's naturalism, it is far from Dewey in spirit, since the practical thrust of Dewey's philosophy has been dropped. Philosophy as therapy is consolation for a world out of control. Pragmatism is the meliorist attempt to make sense of and improve the world. Rorty is a passive materialist; Dewey is not a materialist and decidedly not passive in the face of an indifferent cosmos. The difference between Dewey and Rorty on this score has, ironically, been summed up by another famous materialist philosopher:

"Philosophers have only interpreted the world. The point, however, is to change it."

I

ENVIRONMENTAL ETHICS AND INTRINSIC VALUE

In this chapter, different philosophies containing models of environmental ethics, which are based on some form of the intrinsic value of the nonhuman, will be examined. The authors discussed are three of the more prominent environmental ethicists[1] who base their approach on the intrinsic value of the nonhuman realm to various degrees. These authors will be presented in increasing order of radicalism. Tom Regan argues for the intrinsic value only of higher animals, excluding intrinsic value from plants and lower animals. Thus although he argues for the intrinsic value of the nonhuman, this value is more restricted in scope than in the other two authors. J. Baird Callicott, following Aldo Leopold, argues for the "land ethic," which attributes intrinsic value more holistically to species, habitats, ecosystems, and the like. Intrinsic value is thereby given a larger scope than in Regan and also a different locus of value, in wholes rather than individuals. Finally, Holmes Rolston III argues for the intrinsic value of much the same set as Callicott, but is more radical in his theory of value. Callicott argues from a Humean position of (human) moral sentiments in which intrinsic value is projected to a nonhuman set of members. Rolston, like Regan, argues that value must be completely beyond any human basis but, unlike Regan, has a much larger set, which includes more than just higher animals. It also embraces other species, ecosystems, and the biosphere. Rolston is the most radically ecocentric environmental ethicist of the three and marks the biggest break with modern Western value theory and ethics.

A subsequent section will detail criticisms of non-anthropocentric value theories from within the literature of environmental ethics. Bryan Norton has developed a sustained critique of inherent value theories in environmental ethics, and argues that environmental ethics can be established with anthropocentrically based values.

I will use certain terms frequently in what follows and some may not be clear. I will attempt to define them at the outset. Anthropocentric means any philosophy or theory of value that makes a special case of humans and is oriented toward humans. A theory of value that is anthropocentric bases value in some distinctive human capacity, whether this is a distinctively human species difference or essence, a psychological faculty or capacity, a subjective state of which only humans are capable, or some other factor exclusive to humans. It stands in contrast with animal rights theories, which extend the scope of value to animals; ecocentric theories, which are centered in the environment; and biocentric theories, which are centered in life. Ecocentric means a philosophy or theory of value that is rooted in the ecology. Subjectivity means both those philosophies grounded in the human subject and their theories of value. It is a subset of anthropocentric, as the subject is generally human. However, some have argued that animals are like human subjects in the relevant respects, and differ only in degree. Thus animals can be the subjects of a life of a sort. Subjective theories are usually contrasted with 'objective' theories, which ground value in the object.[2] Relational theories combine subjects and objects by locating value in a relation of an object to a subject.[3] Finally, there are 'foundational' and metaphysical theories of value in which value goes beyond the subject-object distinction and is more primitive, basic, or fundamental.

Intrinsic value is frequently defined idiosyncratically, contextually, or, for some authors, using only a partial definition, at least in comparison to other authors. The dictionary definition of "intrinsic" is "belonging to the essential nature or constitution of a thing."[4] It is sometimes equated with "inherent," that is, "the essential character of something."[5] Other authors distinguish inherent from intrinsic value, notably C. I. Lewis. Intrinsic value is usually contrasted with instrumental, extrinsic, or use value. Often the intrinsic-instrumental distinction is coextensive with the ends-means distinction, but not always. Similarly, it is usually coextensive with the distinction of actions, subjects, or things that are valuable for their own sake by contrast with those valuable for the sake of something else. These terms will be scrutinized in more detail in the book.

The "locus" of value (or rights) is the instance, level, or locale of such value including the type in which value is placed.[6] However, locus of value could include places—for example, habitats and landscapes. It can also include analytically distinct relations (e.g., ends and means), if value is located in one or the other. Nominalistic theories argue that value can only be located in individuals. Holistic theories, by contrast, place value in larger, often abstract loci, such as species, ecosystems, and the biosphere. "Bearers" of value[7] (or rights) are individual instances or tokens of a locus of value (e.g., one individual in a nominalistic theory, or one species in a more holistic theory). Locus and bearer can be the same (e.g., an individual), but need not be.

"Moral considerability,"[8] like moral standing, means those bearers of value that are entitled to receive moral respect or consideration. Only bearers with moral standing can be due moral responsibility by moral agents. Moral responsibility does not extend to bearers of values without moral standing, for example, purely instrumental values.

Regan's Animal Rights

Tom Regan is perhaps the earliest author of the three we are considering on the subject of intrinsic value in environmental ethics and in many respects he is a pioneer. One of his early articles follows Peter Singer in calling for rights for nonhuman animals.[9] Like Singer, Regan wishes to extend what have previously been considered human rights to nonhuman animals. Unlike Singer, he does not base rights on a utilitarian view but on a novel argument from intrinsic value as the basis of rights. He is critical of the utilitarian view as too anthropocentric and thus as unable to provide a sufficient justification for the rights of nonhuman animals.

Regan argues in "The Nature and Possibility of an Environmental Ethic,"[10] that the traditional ethics of Western philosophy is inadequate to protect animals, species, and the environment. Because it is anthropocentric, grounded in specifically human capacities and attributes, it cannot provide an adequate basis or defense of animal rights or the preservation of species. At best, he argues, anthropocentric ethics can only produce a "management ethic" of how to best manage the balance of nature for human use.[11] It cannot meet what Regan considers to be the first condition for an environmental ethic, which is that "an environmental ethic must hold that there are nonhuman beings which have moral standing."[12] Regan considers a number of what are deemed to be anthropocentric theories and argues that they are inadequate. Regan identifies one such theory as the "contract" view. A contract is a set of rules that humans agree to abide by. As animals cannot understand such contracts, they are excluded from them and thus from moral consideration.[13] Such understanding is another case of a subjective state confined to humans, although it might be argued that it is a unique human ability, rather than a subjective state. Such arguments are, following Singer, "speciesist," in that they confine moral consideration to one species. Similarly, the argument of Kant that mistreating animals may "corrupt" human character[14] is judged inadequate as well as anthropocentric. It fails to grant "moral standing" to animals.

Regan argues that arguments for confining value to human subjects from or based on consciousness are "kinship arguments," which argue that from "the idea that beings resembling humans in the quite fundamental way of being conscious . . . [that they] have moral standing."[15] "Kinship" theories are larger than management theories, that is, the theories that result in a management ethic. They include nonhuman species, but only on the basis of resemblance to humans. Thus kinship theories extend moral standing beyond humans, but are still subtly anthropocentric, as they only grant moral standing to animals that resemble specifically different human capacities. Regan argues that kinship arguments fail to meet a second "condition of an environmental ethic." This condition is that "An environmental ethic must hold that the class of those beings which have moral standing includes but is larger than the class of conscious beings—that is, all conscious beings and some nonconscious beings must be held to have moral standing."[16] Neither the kinship nor the management ethic can constitute a valid ecological ethics as they fail to include some nonconscious beings as morally considerable. The same failing is also a flaw of the many forms of utilitarianism. Although several utilitarians have argued for moral standing for animals, notably Singer, they are not radical enough for Regan. Since they confine moral considerability to "sentient" beings, they exclude a great many species.

Regan contrasts kinship arguments with anthropocentric arguments that exclude animals, even those similar to humans. One of these arguments is what he calls the "interest argument," which states that only conscious beings can have interests and thus moral standing. The interest argument is one variety of the argument from unique human abilities or capacities. Since, as I noted earlier, moral standing is required for moral obligation, there can be no obligation toward those beings who are not conscious in such theories.[17] Regan argues specifically that if what is "benefited or harmed" by what is "given or denied them" is in their interest, it is coextensive with those having moral standing, and then it is larger than the class of those having consciousness. The interest argument is similar with what he calls the "sentience argument," that only beings that are sentient—capable of feeling pleasure or pain—are morally considerable.[18] This is an argument from subjective human states, as sentience is a state of human subjects. Regan argues that exclusive human sentience is not self-evident and fails to provide an adequate justification for preservation of nonsentient species. Thus even if it is valid, it fails to provide a basis for preserving the vast majority of species.

Another argument he considers is what he calls the "goodness argument," that is, that the only beings that can have moral standing are those that can have a "good of their own." As only conscious beings can have a good of their own, only conscious beings have moral standing. This is not a repetition of the "interest" argument, although it also involves consciousness, as "good of their own" is distinct from "interests." The former seems to be equated with either taking an interest in something, if narrowly defined, or, as Regan has argued, being benefited or harmed by something in the wide sense. "Good of their own," by contrast, is connected with "a certain kind of good of one's own, happiness."[19] While conceding that only conscious beings are happy, Regan argues that it is disputable that this is "the only kind of good or value a given X can have in its own right." Be this as it may, the thrust of Regan's argument for a distinctly environmental ethics is that less inclusive ethics provide no rationale, justification, or warrant to preserve nonhuman species. Although "kinship" arguments provide for the moral standing and thus the possible preservation of some species, they do so only by analogy with human capacities or specific differences. The goal of a properly environmental ethics is to argue for the preservation of the environment, including other species *apart from their resemblance, utility, benefit, recreational value, or other instrumental value to humans.*

Regan argues that moral standing requires that nonhuman species "have" intrinsic value. His argument parallels the arguments for moral standing for humans. The arguments that excluded animals from moral standing in the tradition argued that various human capacities gave humans intrinsic value. Intrinsic value was a warrant, justification, or basis for moral standing. Moral standing requires intrinsic value, not vice versa. Thus if the class of those with intrinsic value can be extended, so can those with moral standing. If animals, other species, and the environment can be shown to have intrinsic value, they would be entitled to the same moral standing as humans.

Regan tries to avoid the metaphysical problems involved in attributing inherent value[20] by stating that nonhuman species "have" inherent value. The only other reference to the "ontological" basis of inherent value is what he calls the "presence" of inherent value.[21] A value bearer, then, 'has' intrinsic value and it is 'present' in a locus of value. As Regan is a nominalist, arguing for individual rights on the basis of inherent value,[22] value is 'present' in individual bearers. That is, the locus of inherent value is in individuals. The value of more holistic and abstract loci, such as species, is derivative from this. There does seem to be an emergent aspect of inherent value. Regan states that "the presence of inherent value in a natural object is a consequence of its possessing those other properties which it happens to possess."[23] Because inherent value is a "consequence," it follows on or is derivative from the other properties of an individual. Presumably these are all the properties, not some few, as these conjointly make up the individual.

Regan's view also constitutes a form of naturalistic value theory, that is, that value can be accounted for and derived from, in this case, the nature of the individual as its properties. The properties of the object possessing inherent value are "natural"; value is derived from nature. This is to derive inherent value from what is, an implicit challenge to the arguments of nonnaturalism presented by Moore.[24] However, Regan also argues that inherent value cannot be reduced to the other natural properties. Since it is a "consequence" of possessing the other natural properties, it does not seem to be identical with any one of them or all of them taken as a whole. Indeed, Regan describes inherent value as an "objective property": "The inherent value of a natural object is an objective property of that object."[25] As it is an objective property, it does not belong to it by virtue of human subjectivity or human stipulation. Regan does not state what property this is, or how to describe it, or its relation to the other properties, except as emergent.

As Regan's thinking emerged in the context of "rights," it may be that inherent value is inherent in being individual and is "objective" in a moral sense of belonging to any morally considerable individual inherently. Inherent value attaches to moral beings just as rights do. The evidence for this is that inherent value is assigned to bearers equally.[26] There are no degrees of value, based on degrees of some other property. Equality is a norm that generally arises in the context of morals although it could arise in any theory of universals—that is, the universal attribution of a concept assigns it equally to all instances. Thus it applies to all individuals equally. Inherent values are "objective" as "logically independently of whether (someone) is valued by anyone else."[27] This does indeed define what inherent refers to and is almost a tautology. It establishes that inherent value is independent of human valuing. However, it does not clarify in what sense inherent value is an objective property. Regan avoids the "metaphysical" or "ontological" problem of inherent value here by defining inherent value in opposition to any relation to a valuer: inherent value does not consist in any relation to a valuer. This is an implicit counterargument to the long line of those who have argued that value requires a valuer.[28] If a value is inherent, it is independent of a valuer, specifically a subjective valuer. Because it is independent of any subjective valuer, as inherent, it is 'objective' in some respect—it belongs to the object by itself. This may mean morally objective in the sense I explicated earlier, although Regan does not clarify this point. It best exemplifies the sense of a bearer "having" value as it has value independent of a valuer and by itself. The bearer's having value is objective, as independent of subjective acknowledgment. An inherent bearer has value by itself, apart from valuation by a subject.

Regan also argues that if something has inherent value, it is "not exclusively instrumental."[29] Inherent value is the property of not being an instrument for someone's use, not being solely a means. The reference to ends in relation to means is similar to the third formulation of Kant's Categorical Imperative. The advantage is that the metaphysical problem is avoided: inherent value is not defined in relation to the "objective properties" of things, but rather in relation to the other main division of value, instrumentality. As the latter is relatively noncontroversial, defining it in this relation clarifies its value status. However, the problem is that its "objective property" and "natural" status are not clarified.

One article does describe "inherent value" of humans, however. Humans have inherent value as they are "not only . . . alive, they *have a life*" and are the "subjects of a life."³⁰ Regan argues that this is shared with animals in morally relevant ways, that is, that animals are subjects of a life that is valuable to them. The sense of having a life is both owning one's life and being aware in some sense of this ownership, and seems to be a consequence of "natural" properties that are more than any one property of an individual. Being the subject of a life is consequent on having a life. Both being the subject of a life and having a life are objective in the sense that they can be confirmed, at least for humans, and remain independent of whether or not anyone else values them.

I noted earlier that moral considerability depends on inherent value. Reagan argues both that anything with inherent value is entitled to "respect" and "admiration," and also that it should not be treated as a "mere means."³¹ "Respect" is a word that is also prominent in Kantian ethics; not being treated as a means summarizes the third statement of the Categorical Imperative. Thus Reagan seems to be arguing that although he rejects Kant's limitation of moral standing to humans, the argument from inherent value of an end to moral standing is valid. The difference lies in the extension of moral standing to nonhumans based on relevant likenesses. Those with moral standing should not be treated as mere means since they are entitled to respect and admiration, that is, to be treated as ends in themselves. The crucial difference is whether nonhumans have inherent value. If so, they are entitled to moral considerability. Regan makes several arguments for the likeness of animals to humans in respect of morally relevant value. One is that as in humans, death "forecloses satisfactions." As with humans, animals act in the present to bring about satisfactions of their desires in the future, for example, in foraging or hunting for food.³² Just as the death of a young human is tragic because it forecloses potential satisfactions that the human might have experienced had he or she lived, so is it tragic in the case of animals. For Regan, the presence in animals of satisfactions, acting to bring them about, some cognizance of the future, and the continuity of a life are shared with humans, and are grounds for moral considerability. I have already mentioned being the subject of a life as another.³³

Regan advances an argument for extension of inherent value to nonhumans which is different from such normative arguments. By normative I mean an argument that animals and humans normally share certain characteristics, traits, psychological states and processes, or the like, such as having satisfactions. The other arguments are what Callicott has called "arguments from marginal cases."³⁴ These argue that animals differ in no morally relevant way from marginal cases of humans, those who are comatose, retarded, immature, or otherwise "abnormal," marginal cases. To the argument that animals cannot articulate their own interests, or practice duties, it is argued that many humans outside the norm cannot either, but are not excluded from bearing rights on such a basis. Thus such normal human characteristics should not be used to exclude animals. Regan explicitly considers sentience, having a "good of their own," ability to recognize and follow a contract, having interests, and having feelings, in making the arguments from marginal cases. Different authors have advanced all these as grounds for excluding nonhumans from moral considerability.[35] Regan argues that such grounds would exclude a great many humans as well, in morally unacceptable ways. Human babies cannot keep contracts, for example, but are still extended moral considerability.

If something has inherent or intrinsic value, it is entitled to moral considerability. This is the hidden minor in many of the intrinsic value arguments in environmental ethics. The conclusion, that moral agents have a duty to protect bearers of intrinsic value, does not follow directly from the "presence" of intrinsic value. The minor is required to connect value to obligation. This creates a warrant, ground, reason, or justification for the protection of nonhuman nature, however the latter is defined. Environmental ethics, then, is within the rationalist tradition of the West in attempting to justify its ethical mandates with reasons. It does not make irrational appeals. Nor, in view of its appeal to reasons, is it as much of a break with Western philosophy or ethics as some have claimed or might prefer,[36] at least in form—that is, in appealing to reasons and justifying ethical imperatives with a warranted ground of some sort.

What is interesting is that this has taken the form of an axiological ethic. An axiological ethic is an ethic based on values,[37] not nature, the subject, and so on. Duty is derived from value or has a necessary relation to value, which is at least somewhat striking in light of the topic of the environment, where appeals to nature rather than value might seem apropos. Moreover, the notion of intrinsic value is a controversial one and the bearer of such value is a major issue of dispute within value theory. Thus intrinsic value might appear a shaky premise on which to build an ethics. Further, value theory is itself a field of dispute with subjective, objective, relational, and foundational theories to speak nothing of varieties of these. Because the value problem has not come to the point of consensus, building an axiological environmental ethic might seem to be an ambitious undertaking and dangerous to the project of protecting the nonhuman realm. To posit the value of nonhuman species is to commit oneself to a value theory and if such a theory is itself proven fallacious or invalid on other grounds, the basis of the whole ethic is threatened.

The source of such arguments seems to lie in Kant and Bentham. Bentham, particularly, suggested that the extension of moral standing has increased over historical time and may yet expand to include animals.[38] Kant argues that moral standing only extends to humans;[39] however, he agrees with the crucial premise that treating someone as an end (i.e., as a bearer of intrinsic value), involves moral considerability, and thus a duty toward such a bearer.[40] Thus although Kant rejects the notion of the moral standing of animals, differs in his value theory from Bentham and even in the relation of value to obligation, he agrees that intrinsic value entails duties, or at least that there is a rational relation between intrinsic values and duties. Thus the extension of moral considerability to animals, if it could be shown to be warranted, can use the Kantian-Bentham arguments to extract duties to nonhumans. This is the course the non-anthropological ethicists have taken. The premise is hidden in that although many authors who argue for extension of intrinsic value to nonhumans have many arguments for extending intrinsic value and thus moral considerability to nonhumans, they almost never consider the premise that intrinsic value entitles one to moral standing. This could be taken to mean that they have a moral theory of intrinsic value. However, there are nonmoral theories of intrinsic value, notably esthetic theories.[41] For esthetic theories of intrinsic value, intrinsic value does not necessarily entail moral standing. Thus it would seem that the attribution of intrinsic or inherent value to nonhuman species would require a defense of the premise that intrinsic value entitles the bearer to moral considerability, or a defense of a moral theory of intrinsic value. Regan does not provide this, perhaps because in the context of morals it is not perceived as relevant. Again, the defense of moral intrinsic value by Kant and others is perhaps perceived as sufficient justification of the connection between inherent value and moral considerability. The environmental ethicist only has to justify the extension of moral considerability to nonhumans, not the premise of the moral standing of bearers of inherent value.

What is even more radical in their position is that there has been no attempt to go beyond axiological ethics by grounding value in epistemology, metaphysics, ontology, the subject,[42] or any of the other traditional philosophical foundations. In effect, values are their own foundation. This may follow from the notion of "intrinsic" value, as an intrinsic value does not need reference to a subject, or a further ground. It is self grounded, an end in itself. However, the issue of the grounding of intrinsic value is sometimes not raised by the environmental ethicists. The grounding is taken as a premise and is generally assumed. As I noted earlier, it may have been considered unnecessary to do so as the argument for moral considerability was derived from various traditional, anthropocentric ethics—namely, those of Kant and Bentham. Since these theories are well established, it may have been thought that no further justification was needed. If moral considerability was sufficient to justify duty to humans, then the only task needed was to extend this to animals, in order to provide protection to them as well. Justifying the entire ethical enterprise or even the argument from intrinsic value to duty is unnecessary so long as it is not challenged for humans. However, a problem may be created for an environmental ethics that goes beyond the locus of value in individuals in the original theories—that is, Callicott's and Rolston's location of intrinsic value in wholes, such as species, habitats, and ecosystems, rather than in individual value bearers. The original premise was that individuals were bearers. Does this shift in the locus of value change the premise of the argument as originally formulated, that is, that bearing intrinsic value entitles the bearer to moral considerability? This point will have to be considered in a later section, when Callicott's and Rolston's theories are being examined. In any case, it does not apply to Regan, as the locus of value in his theory is clearly in individuals, both as having intrinsic values and as having rights.

The final consideration is the moral obligation due to the bearers of moral standing or considerability. Regan argues directly from inherent value to moral obligation and also to rights, skipping the minor. These conclusions are distinct, but the premises and form of the arguments are the same, that inherent value entitles the bearer to rights or generates duties from moral agents. What Regan calls the "rights view" of the correct relation of humans to nonhumans is stressed in his earlier essays, for example, "The Case for Animal Rights," where he states that all who have inherent value have "an equal right to be treated with respect."[43] Later, perhaps under the influence of critics of the rights view by others, for example, Callicott,[44] Regan stresses obligations as a conclusion. In general, Regan argues that such obligations consist in a "preservation principle" as a "moral imperative,"[45] that is, an obligation to preserve both individuals and species. The equality of rights is held over in the form of equal inherent value. Regan argues against degrees of inherent value, thus degrees of obligation. "All who have inherent value have it equally, whether they be human animals or not."[46] Other obligations include the elimination of laboratory experiments on animals.[47]

The derivation of the conclusion from the minor is not explicit in Regan and may be borrowed direct from previous moral theory. Moral considerability is seen as entailing moral obligations or rights. While this is seemingly plausible, it does not give a principle of selection. If there is a conflict of rights or of obligations either between humans and nonhumans, no principle of preference is given, although Regan argues on pragmatic grounds that humans should receive preference. Further, there is no principle for choosing in the case of selecting between species. This creates a considerable problem for Regan's and other theories in which the locus of value lies in individuals. As Callicott has noted,[48] it means that carnivores and other predators cannot take lives for food because they might be violating rights. Further, it might be argued that humans have an obligation to prevent predators from killing for food, as moral agents. Because Regan argues for an unspecified "preservation principle," it might seem that only preservation of species is an obligation. However, this would be directly contrary to the argument that bearers of inherent value are morally considerable and that rights attach to individuals. This is, it seems to me, a major problem with attempts to locate rights or inherent value in nonhumans, however defined.

The way around such a dilemma is threefold. One can question the location of intrinsic value in individuals. This is the strategy of Callicott and

Rolston, who locate value in wholes, whether species, ecosystems, "the land" or biotic communities. Another strategy is to question whether inherent value automatically entitles the bearer to moral considerability. Rolston questions this minor premise as well. A third strategy is to grant moral considerability, but deny that obligations are coextensive with the bearers of moral standing, and argue for a graduated scale of obligations. Callicott urges this approach.

Another criticism of Regan's theory from within the camp of those who argue for the intrinsic value of the nonhuman[49] has been that Regan's theory does not provide protection for species, ecosystems, biotic communities, and other more abstract ecological formations.[50] Thus species of plants might go extinct and never be defended by Regan's theory. More, the habitats of even the higher animals might be at risk, to speak nothing of attractive esthetic landscapes, wild rivers, and other exotic locales. Regan only defends higher animals as morally considerable, not plants and other candidates. Since one goal of an environmental ethic even as stated by Regan is to successfully argue for the preservation of species and perhaps their habitats, Regan's theory is as inadequate as the anthropocentric theories in achieving this goal. In the next two sections I will consider authors who attempt to extend intrinsic value beyond higher animals to the "land" and biotic communities.

In this section the discussion of intrinsic value has already unearthed several relations "inherent" to intrinsic value arguments. One is the relation of instrumentality to inherent or intrinsic value used by Regan to distinguish inherently valuable bearers from instrumental value. As this characterization of the relation does not refer beyond the value dimension, I will call this the relational aspect. Another is the correspondence of the instrumental-intrinsic value distinction to the means-end distinction. Regan refers to this correspondence in consideration of treating bearers of intrinsic value as ends only. As an end is teleological, I will refer to this as the teleological dimension of intrinsic value. The assumption is that the instrumental-intrinsic value distinction is either identical with or coextensive with that of means to ends. Otherwise, if the means-ends relation is not coextensive with the instrument-intrinsic relation, these are not aspects, but independent dimensions. A third relation is if something is considered valuable "for its own sake," rather than "for the sake of" something else. Another way in which this relation can be expressed is value "for (something else)" and value "for itself" or "in itself" or "in and for itself." Regan refers to this distinction in his treatment of subjects of a life. I will refer to this as the reflexive aspect or dimension. Reflexivity also seems to parallel the relational aspect of instrumentality to intrinsic value, but is not identical with it; something may be valuable "in itself" without having any relation to an instrument. Thus it is a distinct aspect of intrinsic value. These aspects do not necessarily coincide with either the locus or the bearer of value. They are aspects of value apart from any relation to bearers. The bearers' relations to other bearers may not coincide with any of these value relations. This will prove to be an important point in discussing value loci in Callicott and Regan.

Callicott's Biotic Community

J. Baird Callicott has, like Reagan, called for a new, distinctive, environmental ethic. He bases the need for a new ethic on the inadequacies of traditional Western ethics for protecting the environment. More, Western philosophy has been unable to envision the intrinsic value of the environment and to deal with it on its own terms. Callicott has not only called for a radically new environmental ethic, which breaks with Western ethics in several important respects, but also for a restructuring of philosophy itself as part of this program. Philosophy is seen as captive and a new approach is required to bring it more in line with both current scientific thinking and the "land ethic."

Callicott argues that, "since Western moral philosophy has been overwhelmingly if not entirely anthropocentric—i.e. focussed exclusively on human welfare and the intrinsic value of human beings . . . the environment enters into ethics, upon such an approach to environmental ethics, only as the arena of human interaction. The environment is treated as . . . a value neutral vector. . . ."[51] Anthropocentric moral philosophy, it is argued, only acknowledges the intrinsic value of humans.[52] "An anthropocentric value theory, by common consensus, confers intrinsic value on human beings and regards all other things, including all other forms of life, as being only instrumentally valuable. . . ."[53] Intrinsic value is immediately introduced as a central topic or issue in environmental ethics. This statement of the problem introduces the relational aspect of intrinsic value to instrumental value within the context of anthropocentric and nonanthropocentric ethics. Because anthropocentric ethics only confers intrinsic value on humans, and only instrumental value on the balance of the world, it is judged inadequate as an environmental ethics. The implication is that an adequate environmental ethic requires that the nonhuman sphere, in some sense to be defined, is recognized as having intrinsic value. Without such recognition, no environmental ethic can be adequate. Treating animals or other elements of nature as instruments is not only to treat them as inferior, but also as outside the sphere of moral consideration. In this point Callicott essentially agrees with Regan. Effects of human policies on the nonhuman realm are considered, if at all, only indirectly.

Callicott recognizes the importance of metaphysics and systematic frameworks in shaping value theory. In particular, he has argued that the Cartesian metaphysics with its framework of a conscious subject confronting a value neutral object has been at the root of the devaluation of the nonhuman, including animals, plants, habitats, landscapes, and their relationships.⁵⁴ Descartes not only argued that nonhuman nature was mechanical, soulless, and without consciousness,⁵⁵ he initiated the dualism of subject and object, mind and matter, man and nature that split value from world and confined it to the human subject. Dualism effectively gave a metaphysical basis for egoism as the solitary subject was identified as an ego, and its worldview is necessarily egoistic. Egoism has been taken as self-justifying by modern ethics in the form that self-interest is defined as "rational."⁵⁶ The modern form of anthropocentric ethics is egoistic, having regard only for the good of human selves, not the common good or that of the nonhuman.

An anthropocentric ethic, with a metaphysical justification of egos detached from the natural world, can disregard the value of the environment, except perhaps prudentially.⁵⁷ Callicott labels such a prudential ethic "utilitarian," in the sense that it regards the nonhuman as a field of utility or use value, that is, as only having instrumental value. Thus there is an equation of instrumental and use value and the implication that intrinsic value, which has been contrasted with instrumental value, is non-utilitarian. "The deeper philosophical problem of the value of the natural environment in its own right and our duties, if any, to nature itself was ignored."⁵⁸ Callicott contrasts this disregard with the attitudes of non-Western cultures, especially those of Native Americans, who, he argues, put a much higher value on the nonhuman, often treating other species as tribes or societies in their own right.⁵⁹ Callicott argues that in view of disappearing wildlife, rapid extinction of species, loss of habitats, and the like, the problem of developing a non-anthropocentric value theory is "the most important philosophical task for environmental ethics."⁶⁰ While Callicott acknowledges that species have gone extinct in the past, as part of a cycle that occurred long before humans arrived on the scene, he argues that the abrupt and catastrophic extinctions caused by human expansion into the wild and relentless exploitation of the environment is something new and unprecedented. This has not only resulted in biological impoverishment. There is something wrong with the wanton destruction of wildlife that goes beyond human loss or concern. Nothing less than a "paradigm shift" in moral philosophy is required.

A new environmental ethics that includes the wilderness is needed, and that is more than a reapplication of older theories, an extension of traditional anthropocentric ethics.⁶¹ Callicott prefers what he calls an "ecocentric" approach, which, instead of starting from the environment and looking for some suitable theory, aims at the complete overhaul of Western philosophy. This will be based on "a shift in the locus of intrinsic value from individuals (whether individual human beings or individual higher . . . animals) to terrestrial nature—the ecosystem—as a whole."⁶² He wants to provide a reasoned non-utilitarian justification for the right of other species to exist, based on their intrinsic value. The shift to species as the locus of value marks his break with less radical environmental ethics, which he regards as still anthropocentric in some respect. This point will be covered later; at this point its focus as a point of difference with both traditional anthropocentric ethics and certain less radical approaches to environmental ethics can be noted.

Callicott claims that the science of ecology has established a new view of the environment that has to be taken into account in an environmental ethic. The early mechanical view of the interrelation of niches and species has been replaced by a view that stresses energy flows, food chain relations, and communities.[63] A "complicated web of relations" is involved in any environment that determines the interaction of organisms. An individual is continuous with the web and constituted by it.[64] The role of species in the whole outweighs the importance of any one individual. An individual may be killed off as prey and the species survive to play its continuing role in the whole. Callicott lays major stress on the role of species in the whole—that is, holism as opposed to individualism, as part of his thrust toward a reformed ethic of the environment.

Callicott is attempting to establish a "foundation" for a new, ecocentric ethics in "an evolutionary and ecological understanding of nature. . . ."[65] He argues that the new sciences collectively studying the environment have radically shifted the paradigm of how nature is to be understood, and the human place in it. In turn a new, environmental ethics based on the understanding of "biotic communities" is called for. "The twentieth century discovery of a biotic community has helped us realize the need . . . for an environmental ethic." Just how radical a break is such an ethic with traditional, "anthropocentric" ethics? Is an environmental ethics an ethics at all from the traditional point of view? Callicott argues that "an environmental ethic is supposed to govern human relations with nonhuman natural entities."[66] Ethics is to be reformed by moving away from a strict concentration on humans and their relations to include the nonhuman. Humans would be included in the web that ecological sciences have discovered. Callicott's project is more radical than a simple extension of moral considerability to the nonhuman from the human, as in Regan. It is ecocentric, and starts from the ecology, not from the human sphere. However, there is also a notion of expanding intrinsic value from the exclusively anthropocentric view to include the nonhuman. If the detached, egoistic self of the Cartesian view is one with the world, within the ecological web, then the ecology gains an intrinsic value it was not previously thought to have, and the destruction of the environment is perceived as a loss to me.[67]

Callicott conceives of his project as a radical departure not only for ethics but for philosophy as well. His reading of the history of philosophy is that it suffers from "physics envy" and is need of reform "from the ground up." The "new paradigm" is to be based on ecological studies; thus it is less a break with the model of philosophy as based on some relationship with experimental science than a shift in which science is to be used in the relationship. In a sense, this is less of a revolution than a return to philosophy's "dedicated place and role in Western cultural history."[68] Callicott means that philosophy ought "to redefine the world picture ... to inquire what new way we human beings might imagine our place and role in nature; and to figure out how these big new ideas might change our values and realign our sense of duty and obligation."[69] The worldview of the Cartesian subject, with a mind confronting an alien and strictly mechanical world, is replaced with one in which humans are part of an ecological web. The implications are that a new ethic must be formulated in this light for which the relation of human and nonhuman must be at the forefront. Although philosophy should still work with a scientific background, Callicott argues that contemporary science has completely surpassed the worldview represented in Descartes and that philosophy must change accordingly. This entails a new axiology that is neither subjective nor even objective.[70]

Why is Callicott's theory formulated in view of Regan's earlier work along the same lines? Regan already called for a new environmental ethics, based on the intrinsic value of nonhuman animals, their moral standing, and thus extension of rights to animals. Why did Callicott need to articulate his own theory if Regan already covered such ground? How does Callicott's environmental ethics differ from Regan's? Callicott and Regan agree that a new, environmental ethic is needed and that intrinsic value must be extended beyond the human sphere to include the nonhuman. The difference is that from Callicott's point of view, Regan does not go far enough. First, Regan places the locus of value in individuals, whether human or animal, and this is an extension of the modern, egoistic worldview of the subject, which does not start from the newer ecological studies. The implication is that individuals are of greater value than species; Callicott argues that Regan's "conservative" view cannot provide a justification for preservation of species, particularly endangered species, in any conflict. "There is no logical link ... between a concern for the intrinsic value of *individual* plants and animals and a concern for *species* preservation."[71]

Further, the value of the *whole* is not considered. The whole is the biosphere or the ecology taken to include all the factors relevant to life, such as soil, water, air, as well as plants and animals. Callicott's holism is an even more radical step away from Regan, as the value of nonliving formations is being considered, not just living things. Callicott argues that Regan's view is simply an extension of the moral standing of traditional utilitarianism, sentience, to certain higher animals, a view he considers inadequate.[72] Callicott calls Regan's view "humane moralism," that is, the view that humans have certain characteristics deserving of moral consideration that the animal liberationists wish to extend a little so as to include the higher animals. The latter view does not take into account the role of nonliving factors such as soil, which are crucial to a biotic community. Animal liberation does not provide a rationale or justification for protection of species, habitats, or landscapes.

Indeed, in one of his more radical articles, he argues that domesticated animals may well pose a threat to the environment, that culling of wild herds may also be requisite, and thus that the lives of individual animals may be sacrificed for a larger whole, although this view is later modified.[73] Thus from Callicott's perspective, Regan's view does not extend moral standing far enough, and, based as it is on an originally anthropocentric view, fails to consider the value of nonliving factors in the biotic whole. It is not a sufficient enough break with anthropocentrism.

Callicott contrasts Regan's "humane moralism" with the "land ethic" of Aldo Leopold.[74] Callicott adopts the land ethic as his own, and articulates and defends its ramifications.

> The land ethic, founded upon an ecological model of nature emphasizing the contributing roles played by various species in the economy of nature, abandons the "higher/lower" ontological and axiological schema in favor of a functional system of value. The land ethic ... is inclined to establish value distinctions not on the basis of higher and lower orders of being, but on the basis of the importance of organisms, minerals and so on to the biotic community.[75]

The land ethic, far from being anthropocentric, is grounded on ecology, breaks with Western individualism in its "functional system of value," and subsumes the value of individuals to that of the biotic community. For Callicott, environmental ethics is the land ethic, and animal liberation is simply seen as, at best, an inadequate forerunner of the land ethic, which should be superceded, and at worst a problematic view that may get in the way of a land ethic. In an early text, Callicott argues that this is because, above all, animal liberation is "atomistic or distributive in ... theory of moral value ... [whereas] environmental ethics is holistic or collective."[76] This shift marks a change both in the locus of intrinsic value, in species, ecosystems and nonliving factors such as soils, and so on, but also in moral considerability. This view gives preference to the land or the biotic community as a whole over individual organisms. The latter are considered with respect to the function their species plays in the biosphere or a particular habitat taken as a whole. This view has created considerable controversy, as individual rights among humans is the consensus view in ethics, and Callicott's view seems to undercut individual rights. However, Callicott later clarified his view and argued that he was not attempting to undermine individual rights for humans.

Callicott speaks at various times of the "land ethic" the "biocentric value orientation of ethical environmentalism," and of ethical "holism" in a way that indicates that these are interchangeable terms. However, in one reading, "biocentric" could be taken as taking all of life as intrinsically valuable. Callicott goes to considerable lengths to separate himself from this latter view. Thus what he means by biocentric must be read in the light of his holism. For the land ethic, domestic animals can be a blight. More, in Callicott's view both animal liberation and reverence for life views are unrealistic in denying the food chain. As animals higher in the chain necessarily feed off those lower in the chain, it is impossible to grant rights in the individualistic sense or intrinsic value to *all* of life. However, Callicott also argues that rights may be recognized within a community—namely, a human community. His point remains that the extension of moral considerability based on the land ethic is not at the same time an extension of individual rights to all living organisms. For this would be incompatible with both the food chain and the health of the land.

Callicott agrees with Regan that an adequate environmental ethic must include the intrinsic value of the nonhuman, but argues that this is a larger set of members than just individual animals. "An adequate value theory for non-anthropocentric environmental ethics must provide for the intrinsic value of both individual organisms and a hierarchy of super-organismic entities—populations, species, biocoenoses, biomes, and the biosphere."[77] "The intrinsic value of our *present* ecosystem as well"[78] must be included in the theory. This would constitute a non-anthropocentric theory of value bearers as it would "provide for the *intrinsic* value of nonhuman natural entities."[79] The value of the nonhuman, as broadly defined by Callicott, cannot be reduced to instrumentality for humans, whether "our interests or our tastes."[80] In other words, the intrinsic *value* of nonhuman individuals is recognized, but not their "rights."

Callicott accepts the three aspects of intrinsic value relation present in Regan, that is, the relation of intrinsic to instrumental,[81] of means to "ends in themselves" or the teleological aspect,[82] and the reflective aspect of value "in and for itself" not for something else. "Something is intrinsically valuable if it is valuable in and for itself—if its value is not derived from its utility, but is independent of any use or function it may have in relation to something or someone else."[83] Because Callicott makes no attempt to distinguish these three aspects ontologically or as moments of a process, they are overlapping aspects rather than distinct relations—that is, the means-end relation is coextensive with the reflexive and relational aspects of intrinsic values. However, he will later modify the character of the reflexive aspect. These three aspects cover the two distinctive "kinds" of value in general: intrinsic and instrumental. The ontology of intrinsic value is more explicit in Callicott than in Regan; "entities" are said to "be" intrinsically valuable and a thing "is" intrinsically valuable. The way in which such entities and things are intrinsically valuable will be covered later.

Oddly, however, Callicott cannot quite break with the Cartesian legacy. For his theory of intrinsic value is ultimately grounded in the subject, that is, in consciousness. This point will be covered in detail later, but Callicott is not as radical in his intrinsic value theory as in his theory of the locus of value and of moral considerability. Basically, Callicott presents a theory of intrinsic value that is subjectively grounded, and in which intrinsic value is projected on to nonhuman nature. Thus something nonhuman that is intrinsically valuable is both valued by someone and "valued for itself." This view meets the criteria for what several authors have called a "relational" subjective view of intrinsic value.[84] Value is "intrinsic" in the object but this is grounded in a valuing subject. Callicott has a less radical view than that of Regan on this point, as the latter argues for inherent value without regard to a valuing subject, despite Callicott's approval of Regan's definition of inherent value, including Regan's qualification that inherent value "must be objective and independent of any valuing consciousness."[85] The "objectivity" of intrinsic value is stressed in Callicott's own analysis. Clearly, this is a complex notion of intrinsic value. Value seems to be grounded in a subject but independent of the subject.

Callicott's position is close to that of C. I. Lewis, despite Callicott's disclaimer. Distancing himself from both Regan and Lewis, Callicott chooses to defend the "intrinsic" as opposed to the "inherent" value of the nonhuman. As he conceives it, intrinsic value is "objective" and "independent of all valuing consciousness," while inherent value is "not independent of all valuing consciousness" even if it valued for itself and not only as a means.[86] In other words, Regan's "inherent" value is included in Callicott's "intrinsic" value. It is not completely clear why Callicott prefers "intrinsic" to "inherent." It could be that Callicott wants to make a stronger case against consciousness as some sort of criteria for intrinsic value or moral standing. Soil and air are not conscious entities; Regan's criteria of sentience involve some sort of consciousness, even of nonhuman animals. Since Callicott is familiar with Lewis's value theory, it could perhaps be that he is basing this distinction and arguing his position in reaction to Lewis. In fact, Lewis uses just the opposite terminology: intrinsic is tied to the conscious subject while inherent is objective. Callicott argues that his position is stronger than that of Lewis, as Lewis's theory is "actually instrumental" with regard to the nonhuman, and that his own theory recognizes the value of the nonhuman for itself.[87] Callicott's position on the ground of value is nevertheless similar to Lewis's as both distinguish subjective and objective value and ground the latter in the former.

Callicott is less than clear on how value can be "independent of consciousness" but grounded in consciousness. However, as ground involves a relation to a consequent, perhaps objective intrinsic value is posited as a consequent. This is one of two solutions proposed by him for the relational grounding of intrinsic value. On the one hand, Callicott agrees with the modernists that there is "no value without an evaluator."[88] Thus subjectivity must ground value, even intrinsic value defined as independent of consciousness. Noting that the problem of intrinsic value is "frankly metaphysical," Callicott argues that "we need to discover . . . metaphysical foundations for the intrinsic value of other species."[89] He argues against any naturalistic approach, perhaps having Rolston in mind, apart from some "valuational consciousness."

Realizing his predicament, that he wants to both maintain the intrinsic value of the nonhuman apart from consciousness and also maintain subjective grounding, Callicott casts about for a way out of his dilemma.

> If intrinsic value cannot be logically equated with some objective natural property or set of properties of an entity independently of any reference to a subjective or conscious preference for that property . . . the only way to rescue the objectivity and independence of intrinsic value is desperately metaphysical.[90]

The "desperately metaphysical" way of rescuing intrinsic value is "to commend a property to our evaluative faculty of judgment or our evaluative faculties."[91] This seems to suggest that a value judgment can be made that nonhuman entities are intrinsically valuable. Callicott is unwilling to give up consciousness as the source of value as it is "institutionalized" in the scientific worldview, that is, in the "Cartesian framework" of subject and object involved in "value-free" descriptions of the natural world.[92] "I concede that, from the point of view of scientific naturalism, the *source* of all value is human consciousness, but it by no means follows that the *locus* of all value is consciousness itself or a mode of consciousness like reason, pleasure or knowledge."[93] This judgment involves the distinction of the locus of intrinsic value from its source. Thus value flows from a source to a locus in a relation. We judge something to have intrinsic value independently of ourselves. Callicott's view does not escape a relation to a subject, however, although it may establish intrinsic value outside of a valuing subject.

Callicott's other solution to the problem of a relational theory of intrinsic value is more speculative. He argues that the subject-object distinction of Descartes as a framework makes the axiological dichotomy of fact and value "intractable." But he also notes that in the new physics, the subject-object, primary-secondary quality and essence-accident distinctions are entirely superceded: the observer and observed cannot be entirely separated. All qualities are secondary in this view, "potentialities which are actualized in relation to us."[94] No properties are intrinsic, even those established by science, that is, ontologically objective and independent of consciousness. But nature can still be valued for its own sake since this does not change the relation of subjectively grounded valuations of intrinsically valuable loci of value. Values would in this case be the same as other properties, "actualized upon interaction with consciousness."[95] Although this solution may preserve intrinsic value, it undermines the need for a Cartesian framework, and thus the subjective-objective distinction on which Callicott's theory is based. Thus it may undermine the whole problematic that gave rise to Callicott's value theory and require a new theory of value, including intrinsic value. Be this as it may, Callicott makes it clear that this is undeveloped speculation.

Callicott also considers other views of intrinsic value and judges them inadequate. The sentience[96] and interest[97] theories of value, even if extended to many lower animals, do not protect species, habitats, or biotic communities as a whole. They are inadequate in scope or the extent of moral standing. Similarly, Kant and Aristotle only extend moral consideration to humans, based on the exclusive intrinsic value of certain human capacities, for example, reason.[98] In general, Callicott is critical of theories that argue that exclusively human psychological states are intrinsically valuable or can be used as a basis for the intrinsic value of the environment. Such theories posit a "hierarchy of beings determined by psychological complexity," whether this involves degrees of rationality, sentience, or desire.[99] Callicott judges such speciesist distinctions "arbitrary" as the basis for the intrinsic value of humans; he asks why they are to be considered good and thereby a basis for intrinsic value.[100]

Moore's theory of intrinsic value is judged inadequate not as limited in scope but because it appeals to intuition, begs the question, and cannot be used to decide controversial cases.[101] The most favorable theory is that of Plato, which Callicott describes as a "holistic rationalism." Plato's view locates the source of value outside the subject in the Form of the Good, and confers values on unified wholes and the harmonic relation of their parts,[102] precisely what Callicott is seeking. Unfortunately, Callicott argues, this view does not ground this particular whole, the present biosphere, but any well-ordered whole. Thus although it comes close, it cannot be used as an environmental ethic because it does not provide a sufficient basis for protection of present species.[103]

In place of such traditional theories, Callicott proposes to adopt Hume's moral sentiments theory as a "basis" for the intrinsic value of the environment: to ground morality "in feeling or emotion." Callicott notes that most philosophers have ignored Hume's theory, dismissing it as relativistic. However, he makes the case that Hume's theory could meet his test of consistency with value-free natural science and yet provide a basis for the intrinsic value of the biotic community. Hume's distinction of 'is' from 'ought' separates scientific judgments from bias. Value, however, is "projected onto natural objects or events by the subjective feelings of observers."[104] This analysis of Hume is consistent with Callicott's previously mentioned arguments for subjective grounding: subjective feelings are the "basis" of a relation to a valued thing. However, his theory involves ontological complexities compounded by the use of similar language for truth/ fact judgments and value/ought judgments. As Callicott urges that an entity "is" intrinsically valuable, the form of the judgment is identical with that of truth statements. There is the further problem of how intrinsic value can be part of the nature of the thing independently of consciousness if it is based on subjective feelings.

Callicott acknowledges that Hume's moral sentiments theory is not promising as a basis on which to build a theory of the intrinsic value of the nonhuman. For the "hypothesis" of intrinsic value means "that value inheres in natural objects as an intrinsic characteristic, that is, as part of the constitution of things."[105] Grounding in a subject would seem to be a denial of the possibility of such value. However, he argues that the relation to the subject of an intrinsically valuable entity is based on a distinct type of subjectivity. It is not the subjectivity of subjection to an ego, but an other-regarding subject. He notes that "value may be subjective and affective, but it is intentional, not self-referential."[106] The projection of intrinsic value is precisely the recognition of the independent value of the other. The other is accorded a worth that is acknowledged as independent of the ego and its utility. The subject, by intending an object without reference to self, accords it intrinsic character, that is, a character apart from the subject. This otherness is an acknowledgment of its intrinsic value, as its independent existence apart from the subject is acknowledged, and thus its value on its own, apart from the subject. Callicott uses a newborn baby as an example of feelings projected from a subject onto an intrinsically valuable being. A newborn is a source of joy, "even though its value depends, in the last analysis, on human consciousness."[107] Babies are valued "for themselves" even before they can subjectively value by themselves, presuming they are too young to do so until a certain age. "It 'has'—that is, there is conferred or projected upon it, by those who value it for its own sake—something more than instrumental value. . . ."[108] This example is an instance of the argument from "marginal cases" of humans, whose use by the animal liberationists Callicott criticizes. Babies are humans, but cannot subjectively value, so are marginal cases of humans in this respect. Thus Callicott surreptitiously appeals to marginal cases to justify intrinsic value. However, his use of the argument is more restricted because it is used to argue only that cases of projection of intrinsic value within conventional ethics can be derived. It does not base its entire case for intrinsic value of the nonhuman on the argument. The example also leaves the status of "projection" unclear. Is such projection metaphysical, perceptive, logical, or some other type? The few hints he gives are that it is a case of logical judgment, as the entity "is" accorded intrinsic value. But the ground of such judgments is unclear: what sentiments are the bases for such judgments? Which entities should receive them and why?

Callicott wavers slightly on the point of the degree of intrinsic value, based on his subjective grounding. As I noted earlier, Callicott used the reflexive relation of 'value for,' to valuable 'in and for itself.'[109] Whereas he earlier used 'in and for itself,' Callicott at another point distinguishes 'for itself' from 'in itself' on the ground that since value is based in subjective consciousness, intrinsic value cannot entirely be 'in itself.' "An intrinsically valuable thing . . . is valuable *for* its own sake, *for* itself, but it is not valuable *in* itself, that is completely independently of any consciousness. . . ."[110] This wavering on the ontology of intrinsic value may be parallel with his realization that the locus of value may not be confined to individuals, and thus that the ontology of intrinsic value is even more complex. In any case, intrinsic value cannot be separated from a subjective valuer for Callicott. But the difference between his theory and that of other subjective intrinsic value theories is that intrinsic value is not confined to the subject, but is somehow projected outward onto an other of value. Presumably, such projection involves first, the recognition on the part of the subject of the value of the object of intention and, subsequently, the judgment of intrinsic value, that is, that the entity has value for itself, apart from the subject.

Callicott does not analyze the character of moral sentiments as such in Hume's theory. He does, however, note that they are "other oriented" and "intentional." This relation is possible because feelings of affection for the other are altruistic, not egoistic. Altruistic feelings are the basis for projection of intrinsic value. Thus although he acknowledges that Hume's theory is anthropocentric, it is not egoistic, or at least balances egoism with altruism. Altruistic feelings require an other or object: such feelings are "not valued themselves, or even experienced apart from some object which excites them and onto which they are . . . projected."[111]

Callicott holds onto Hume's theory for another, historical reason. In several of his articles he has traced the subsequent development of Hume's theory after Hume through Darwin to Leopold, and thus as a historical root of both evolution and the Land Ethic.[112] There would seem to be a loose basis for the Land Ethic in the naturalism of Darwin combined with Humean altruistic moral sentiments. Moral sentiments are based on feelings altruistically projected onto objects, persons, and actions for Hume. Moral value, like aesthetic value, is in the eye of the beholder. Darwin, in turn, adopted Hume's theory in explaining the apparent anomaly of occasional acts of seeming altruism in animal behavior—for example, the prolonged care of young by parents of certain species. Darwin argued that pity, kindness, benevolence, and the like are adaptive, as they increase the ability of the species to survive. The "social instinct" is increased by selection as more and more individuals with a proclivity to such behavior survive. Egoism is not basic in the view of Hume and Darwin; altruism is "equally primitive."[113] Thus altruism can be considered as a "natural" basis for intrinsic value. Finally, Leopold's Land Ethic, involving the "biotic community," is seen as a development of Darwinian naturalism and specifically the altruism within species. Leopold, in Callicott's view, expanded Hume's moral sentiments theory to include not just species but nature as a whole. Elements of the nonhuman world "own inherent value, that is, to be valued for themselves."[114]

The step from moral sentiments among humans to their extension to animals and then to nature as a whole involves an expansion of the moral sentiments view to the nonhuman. Animals and other nonhuman entities must be capable of being objects of affection similar in kind to humans, if not in degree. Callicott argues that humans often have had such relations with animals (e.g., as pets) and such contact presumes some sort of bond similar in kind to that between humans. Indeed, domestication is compared to the social contract.[115] Nonhuman species may possess intrinsic value in the "truncated" sense of "for" but not "in itself." Callicott argues that the science of ecology "fosters" such a view, perhaps because the role of organisms in the biotic community is defined apart from human use.

In a sense, Callicott's view of intrinsic value is more conservative than Regan's despite his radicalism on other issues. For he cannot break with the psychological model of value inherited from the modern tradition initiated by Descartes. On the one hand, nonhumans "own," or "have" intrinsic value and even "are" intrinsically valuable. On the other, this value is grounded in a relation to a subject. As the latter is essential to intrinsic value, although a novel view of its relation and scope, it does not provide a sufficient basis for an *intrinsic* view of value. For as Callicott himself notes, such an ascription requires that value be independent of the consciousness of an other. Regan argues correctly that inherent value must be independent of conferral of value by a subject or it does not inhere in the bearer. It is not inherent. Callicott's desire to retain a value free natural science prevents him from getting past the is-ought dichotomy; thus value remains subjectively based for him. He fails to recognize that Hume violated his own distinction by basing value on moral sentiments: a fact or 'is'; and that if the nonhuman is accorded intrinsic value, it can hardly be value free in the sense required for such an ideal. In other words, he confuses bias and value. Further, if nonhuman bearers "own" intrinsic value or "have" intrinsic value, they have it apart from any relation—that is, by themselves. This eliminates any relation to a subject.

I noted earlier that Callicott included certain qualities as bearers of intrinsic value. Also, individuals may have it, for example, babies. Callicott, no more than Regan, indicates what qualities, individual or collective, bear intrinsic value. The ontology of value is not clear in Callicott. Because his is a projective theory, however, it may be that he did not consider it necessary to articulate any ontology of value. Entities have value projected onto them, for whatever reason, by a subjective valuer. Their character excites such a sentiment. To some degree this character is tied to the locus of value, a topic I will cover next. That is, value is attached to a certain locus for Callicott, and this has a subjective basis. Thus the crucial relation is to the subject, not the intrinsic character of the entity. However, this violates the meaning of 'intrinsic' as defined by Callicott himself. A change in meaning proposed in the theory may be indicated: that the character of the object excites a feeling in the subject that, in turn, is the basis for the judgment of intrinsic value. If this is correct, the relation is much more reciprocal than in egoistic theories, as the object generates feelings as well as being the recipient of judgments of worth. The intrinsic character is involved in the judgment, but mediated by the affective subject. This may be important in Callicott's theory of the locus of value.

Callicott argues against Regan that the "locus" of value, from an ecological point of view, should not lie in individuals. Nature "does not respect the rights of individuals."[116] One bit of evidence for this is in the food chain, where organisms lower on the chain are eaten by organisms higher on the chain. If individuals are sacrificed, their value is limited. They become instruments of another animal or parts in some larger whole. In the biotic community, eating and being eaten is part of the ethic. Callicott argues that even treating animals as an abused minority, as the animal liberation movement proposes, misses the point, as it is species that should be preserved, and preservation of the latter may necessarily involve the sacrifice of individuals. For a species of predator to survive, for example, certain individual prey must be sacrificed: lions require freshly killed meat. Callicott concludes, "species have intrinsic value while specimens have only instrumental value," although he fudges this issue by also arguing that not "both" individual and also larger entities have intrinsic value.[117] This conclusion has brought Callicott into conflict both with rights theorists within environmental ethics, who tend to be animal liberationists; and ontologists of value, who tend to assign or locate intrinsic value in individual bearers. Often these are the same persons, of course. Callicott's theory of the locus of value has an eye for the actual processes of nature, not simply human rights theory. Indeed, he rejects the latter quite vehemently as both utopian and not consistent with actual natural processes. This point will be discussed later in the section on moral obligation; at this point it is sufficient to note that Callicott consistently rejects the extension of human rights to individual organisms on both ecological and normative grounds. Species should be preserved as the locus of value. This brings up ontological problems, for how can a species have value over and above its individual members? Is not a species a collection of individuals? Callicott considers critics of this more holistic view of locus,[118] and in his early work rejected a nominalistic ontology of value as incompatible with natural processes, although in his later work he argues for the intrinsic value of both individuals and wholes. He argues that a species is not simply a class but a "historical entity," that is, a genetically connected unit that has a temporal dimension in the passing down of inherited characteristics between generations. He notes that "people meaningfully assert all the time that nonindividuals—unions, corporations, states (as in states' rights), nations, sports teams, species and ecosystems—have various rights including moral rights."[119] The location of rights in legal or other entities other than humans argues back to their having intrinsic value, by Regan's own arguments.

Callicott notes that the land ethic of Leopold requires that the intrinsic value of species may outweigh that of individuals—indeed must outweigh them if predator species are to be preserved—and that the biotic community outweighs even that of species. More, he defines a species as its relation to its environment. Thus he can argue that the relation is prior to the individual, that the whole is greater than the sum of its parts. The internal relations of species to habitats are not some metaphysical unity as in Vedanta, but more Hegelian. Systemic relations, including consciousness, are an extension of the environment.[120] He calls for a shift in thinking from egoism to ecocentrism. Modern ethical theory, Callicott stresses, is egocentric: it generalizes from self to others. The self is the locus of intrinsic value. But this locus does not, and Callicott believes cannot, provide moral considerability to wholes. Ecological theory tends to be holistic: it reverses the primacy of objects to that of relations. The whole "shapes" species in niches. Not organisms but the community as a whole is the correct locus of ultimate value from an ecocentric point of view.[121] Callicott approvingly notes that in conservation literature biological diversity and complexity are often judged "good in itself."

Based on this ecocentrism, he "locates ultimate value in the 'biotic community' and assigns differential moral value to the constitutive individuals relatively to that standard."[122] Thus there is a hierarchy of value based on a holistic standard, which allows for clear choices in the case of conflict. The "biotic community" as a whole has ultimate value, followed by species preservation and finally individuals. It is not clear from this hierarchy what the value of other elements mentioned by Leopold—for example, air, soil, and water—is and their relative place in the hierarchy.[123] Perhaps it is as elements in relation to the biotic community and therefore as elements in relation to a larger whole. Callicott argues that nature considered as a "biotic community" constitutes a new paradigm: not a collection of species-objects, but a living whole. The value of nature is thus transformed; the natural environment is "the community to which we belong."[124] Humans do not confront nature as subject to objects but are members of the biotic community. Callicott argues that the whole is greater than the sum of its parts. The rationale for this with respect to the location of intrinsic value in some ecological whole is that "the various parts of the 'biotic community' (individual animals and plants) depend upon one another *economically* so that the system as such acquires distinct characteristics of its own."[125] These relations are the basis for an ethic of "holistic care for the community."

An immediate problem with such a view is whether it is compatible with the nominalism and subjective grounding of the moral sentiments theory. If intrinsic value is grounded in subjective valuers, it is difficult to argue that ultimate value must be located in a larger whole, either ontologically or axiologically. If the valuers prefer their children to the biotic community, it is difficult to argue that the biotic community should have priority. Second, it is not clear why the need to preserve species justifies suspension of individual rights for wild species. If a species consists of individual "tokens," then they must be preserved as individuals for the species to survive. If all individual whooping cranes die, the species dies with them. Nor is it clear why holism justifies doing away with certain individual rights. Third, why is the whole of higher value than species, given the impetus to preserve species? The environment may be an instrument of species, as well as of humans, without being accorded a greater value than living species. Further, what of a species with no important role to play in a habitat? Is its value less because of its inferior role?

This is to speak nothing of the ontological problems with locating value in a whole. These were discussed earlier, with respect to the relational view propounded by Callicott. It is not clear that there is a species apart from individuals, and thus whether a species can be a locus apart from individuals. Callicott may surreptitiously be arguing a utilitarian view: that the greatest good of the greatest number of the species is served by his value theory. If the majority of the species survives, this is better for the whole. However, he does not argue this explicitly, nor does such a utilitarian ethic follow strictly from his land ethic. To locate value in a species in the sense of some surviving over time is still to locate value in individuals qua species members.[126] It simply makes the location random. Again, the value of the whole may consist in a collection of individuals. It is not clear that nonliving elements require any more than instrumental value, and thus why the whole has greater value than the parts. True, the destruction of an entire habitat may wipe out a number of species. However, it is the effect on species that is decisive, not that on rocks, soil, air, and water as such. A rock that is vital to a species may be of instrumental value, but it is difficult to see why any value whatsoever should attach to one that plays no such role. It may perhaps have aesthetic value, but whether this gives it moral standing is another issue.

As I argued in the section on Regan, the moral obligation to protect nonhuman value requires the premise that what has intrinsic value has moral standing or considerability. This is the second basic issue on which environmental ethics is based as a new ethic, distinct from older ethical theories. Environmental ethicists question whether the expansion of the purview of traditional ethics to also include some nonhuman loci is sufficient to protect species and the environment as a whole. Those who call for a new environmental ethic argue that not only do some nonhumans have intrinsic value, but that such value confers moral considerability, or moral standing. Callicott includes animal liberationists and Utilitarians in the former group of those calling for extension of traditional ethics. He places "neo-Kantians," such as Rolston, "self-realization theorists" such as certain "deep ecologists," and those who follow Leopold's "land ethic" such as himself of those calling for a radical reformulation of ethics around environmental issues in the latter.[127] Callicott's position is indeed radical and its radical character is exhibited most clearly in the issue of what should receive moral standing.

Callicott notes that the rights theory of Regan confers rights on individuals, "the subject of a life," with the criteria of self-consciousness and capacity to "believe, desire, conceive the future, entertain goals and act deliberately."[128] Similarly, "conative" theories, widely conceived to cover Schweitzer's "will to live," defines interests in terms of conations and intrinsic value in terms of interests.[129] Because plants have interests, conative theories have a wider scope of moral considerability than Regan's theory. However, these theories still assign value and moral considerability to individuals. Callicott claims that such a view is impossible from the viewpoint of the land ethic, that is, from a view that starts from the ecology, not human ethics. In the wild, individual lives are not ultimate; equal rights for all animals would not allow for any killing—for example, in predation. The preservation of species may involve sacrifice of individuals since this may increase the overall ability of the species to adapt and survive in a Darwinian-ecological sense. A species per se may not be conative[130] and cannot be the bearer of individual rights. The preservation of species higher on the food chain may require sacrifice of individuals lower on the chain. It may also require preservation of species that are not closely related to humans in terms of sentience or intelligence, but whose preservation is essential for the health of the whole. Plants may be morally considerable.[131] An individual of a species that is rare may have more value than one of a common species. While for conativism and animal liberation individual lives are of equal value, "species preservationists set a much higher value, for example, on an individual furbish lousewort . . . than on an individual whitetailed deer, a commonplace mammal."[132] Thus a scale of values and of moral considerability is implied, which is in conflict with the essentially egalitarian standing of inherently valuable bearers in Regan and other value nominalists.

What Leopold's land ethic and individualistic theories do share is the expansion of moral considerability.[133] Callicott repeats the theme common to many environmental ethicists that there has been a considerable expansion of ethical consciousness in our century to include women as well as ethnic and sexual minorities, even if this is not always reflected in practice.[134] As anthropologists have noted, the boundaries of moral consciousness are the same as the perceived boundaries of society. Changes in society bring corresponding changes in ethical boundaries, up to "human rights" as universal and even the extension of human rights to animals by Regan. Callicott argues that the next step is the "land ethic" of the "community" as the "basic concept of ecology," in which the intricate interrelations between elements in the biotic community, fixed roles, and niches are stressed.[135] Coordinate with tracing Leopold's views to Hume is Callicott's presentation of the land ethic as the ultimate outcome of such an expansion in moral considerability, as an expansion of moral sympathy to animals, plants, and even soil, water, and air.

Leopold himself remarked on the historical extension of moral consideration or standing, Callicott notes.[136] The land ethic "enlarges the boundaries of the community" still further. It provides for the intrinsic value of both individuals and larger "entities" such as species. Callicott compares the biosphere or biotic community as a whole to the nation and a species to a tribe: the expansion of ethical standing to other species is similar to the ancient expansion to the tribe, rather than just the individual or family. The expansion to the biotic community is similar to the expansion to the nation as a whole.[137] Ultimately, this does not conflict with human morals, as the health of the biotic community is in the interest of humans.[138]

The question remains whether preservation of species requires either the extension of intrinsic value or of moral considerability to species. Does species preservation require the suspension of individual rights to the extent Callicott claims, even if they cannot be extended to the extent the animal liberationists propose? As a species is embodied in a certain number of individuals, preservation of a fixed minimal number of individuals of a species, however such a minimum is determined, will preserve the species. The question is whether a species exists apart from individuals.[139] Callicott correctly argues that it is more important to preserve a rare species than individuals of a common one, but this may be possible on an individualistic theory. Further, the point involves degree of moral considerability, rather than the bearer or locus, a separate issue that Callicott seems to run together.

The same points apply to holism. While habitat degradation is bad for local species, it is not clear why habitats require *moral* standing. If they are of instrumental value, they require protection, but it does not follow that they require moral standing. Protection of houses on prudential and economic grounds is widespread, without anyone proposing that houses have moral status. This is of course closely related to Callicott's plausible view of soil, air, and water as integral parts of the "biotic community." But their role is as instruments to life, not agents, even in the most far-fetched scenario. Protection of them is prudential as conditions of species, not as having moral standing themselves.

Finally, Callicott argues that value is "humanly conferred but not necessarily homocentric." Intrinsic value is a "projection or objectification" of the sentiment of humans.[140] If this is so, there is nothing to prevent humans from valuing individual nonhumans, whether pets, flowers, or landscapes, or from preferring some species to others—for example, food sources to "weeds." In other words there may be conflicts between the moral sentiments view, which as originally articulated by Hume, at least, was individualistic, and the ecocentric view in which species and interrelations are more important. Callicott does provide a criterion of selection in case of conflict, which I will cover, but only if it involves conflicts between individuals, species and the environment as a whole. It does not cover conflicts between the source of intrinsic value in moral sentiments and its objective, the biotic community. It is clear that Callicott's own sentiments lie with the land, but his value theory may not support such a view. Another conflict might lie between feelings for what Rorty has called the more "cuddly" animals—Bambi—and the requirements of the food chain, where Bambi is lunch for a carnivore.

Leopold's "land ethic" includes an enlarged notion of moral considerability, as I noted previously. Just what is the land ethic? How does it follow from Callicott's Humean-derived notion of intrinsic value? What are its consequences for the environment and human practice? Callicott quotes Leopold's summation of the land ethic: an action or policy is "right if it preserves the integrity, stability and beauty of the biotic community."[141] It is wrong if it doesn't. The land ethic, somewhat misleadingly named, is not literally a land preservation ethic because it aims at the integrity, stability, and harmony of the biotic community. However, these goals are centered on the wise use of land in Leopold's view. Callicott's reading of this perspective fills out the "land ethic." Integrity provides a norm of holism and community, stability, of preservation, and harmony contributes to the intrinsic value of the nonhuman.

Callicott's argument for an ecological ethic to replace traditional anthropocentric ethics is based partly on the premise that ecological ethics must be grounded in the ecology, not human nature. He argues that Leopold's land ethic meets this condition. As evidence, he provides a genealogy from Darwin to Leopold, demonstrating that the land ethic is consistent with and even to a degree derivative from evolutionary theory.[142] Further, it is compatible with the more recent science of ecology, which stresses wholes, relations within the community of various roles and niches, and the importance of preservation. Finally, Callicott even claims that the land ethic rests on "Copernican astronomy," as the view of earth with its biota as unique in the "desert of space" developed out of Copernicus' decentering of earth in the heavens.[143] This uniqueness of life on earth might be the basis for the development of feelings of kinship with fellow creatures in the community of life on the planet.

For Callicott, the "base or root concept" of the land ethic is, following Leopold, "the community concept." This concept follows Darwinian evolutionary theory, in that Darwin saw other-regarding sentiments as adaptive and competitive in the struggle for survival.[144] Such sentiments among humans tie in directly to expansion of moral considerability. Expansion of moral considerability has a biological basis, then, and its extension to even nonhuman nature would further aid in human survival. This is also Leopold's view. Other-regarding sentiments, or altruism, are ultimately communitarian, as expansion of ethical regard from self to others must ultimately end in regard for a larger whole. The largest community of all is the biotic community, which includes all the nonhuman and the human in one large biotic or life-based community. The biotic community is centered in the land as an indicator of the health of the whole as well as the basis for the whole. The land ethic is based on moral communities, not individuals. The role of the individual is less important than that of other elements.

Callicott describes the land ethic as an "organic, internally related, holistic view of nature," which is opposed to the predominantly atomic, mechanical, and dualistic view of modern philosophy.[145] It is organic as life-oriented, although it does not take individual life as the locus of intrinsic value. It is internally related as stressing relations among species, rather than autonomous, atomic individuals. Finally it is holistic both in viewing such relations as those within the food chain in respect to their effect on the whole biotic community and also placing the value of the whole highest in any ethical conflict between the whole and species or individuals, although this priority does not seem to extend to humans if I read him correctly. The "holistic vision of the world" intrinsic to ecology focuses on relationships. The importance of predators in maintaining the health of the whole system of interrelations is that of a check on overly voracious predation of plants, the bottom or foundation of the whole. Relationships are in terms of the whole in this model, and Callicott judges this to be a suitable model. Not individual rights but the effect on the whole is decisive.

The most radical consequence for practice that Callicott draws from this model is the call for a "shrinkage"[146] of the domestic in favor of the wild, as domestication—that is, human encroachment on the biotic community, is reaching catastrophic proportions. This is justified in Callicott's view because a truly ecological ethic is in harmony with the facts of biology. A biological understanding of ethics, based on Darwin and the science of ecology, understands ethics as "a limitation of freedom of action in the struggle for existence."[147] Such altruism is viewed as having a biological advantage for the species and biosphere as a whole, despite its restriction of individual behavior. Indeed, Callicott argues that reason, the frequently cited candidate for the specific difference of human, and thus the basis for a humane ethic, presumes ethics. "We must have become ethical before we became rational."[148] Thus ethics from a biological-ecological viewpoint does not privilege reason. Limitations on human action are biologically consistent.

In cases of conflict, Callicott consistently rates the whole higher than any part or species, and species higher than individuals. The species as the locus of intrinsic value places the species as a whole higher than individuals, where the rights of individuals and that of the species conflict. Thus the ethic is consistently holistic, as the whole biotic community comes before species and the good of the species as a whole comes before that of individuals. Callicott's straightforward holism has raised much conflict over the status of individual rights. Animal liberationists have attacked Callicott's position as undermining the long tradition of individual rights over against majorities, which is the original basis for extension of moral considerability. In effect he undermines extensionism. Callicott has countered that an environmental ethic, even if it involves an extension of moral considerability, must be based on the ecology itself, not human mores. He regards the call for individual rights as a disguised argument for the intrinsic *value* of the nonhuman, as rights of animals are more "expressive" than "substantive."[149] Rights do not actually attach to "real" individuals unless the call for rights reduces to stating "claims."

Callicott regards the talk of rights as unfortunate on several grounds. One is that the notion of the rights of individuals in the wild is unworkable, even apart from the problem of its derivation from human legal systems. Predation necessarily involves the sacrifice of individuals; this is as true of humans as of wild carnivores and herbivores.[150] Another is that the rationale of an ecological ethic is species preservation, even for the animal liberationists who are the main champions of individual rights. Callicott argues that the preservation of predator species would be impossible on such a view, and that the notion of granting rights to species themselves is "conceptually odd," if not contradictory. Rights talk expresses "moral considerability," but it attaches to persons or "at least localizable things."[151] A species is a class or kind. Callicott thus concludes, "the assertion of rights on behalf of species is incoherent" in an ecological context,[152] and is symbolic of extension of intrinsic value. Equality of individuals, a corollary of individual rights, is not weighed toward rarer and more endangered species, or toward species critical for the health of the biotic community as a whole. Callicott also argues that individual rights of humans are not undermined by the land ethic. The land ethic is not a basis for individual rights in nature, or indeed in general. Its emphasis on communities and wholes as morally significant is designed specifically to bypass the problem of predation of individuals, consistent with actual ecological relations and their overall health.

What is the relation of the land ethic to traditional ethics of humans? Callicott sees no ultimate conflict between the land ethic and human well-being as the health of the biotic community contributes to human well-being in the long run, although that is not its justification. The emphasis on community in the land ethic is entirely consistent with human ethics. "*Ethics and society or community are correlative.*"[153] His view of ethics is that it always includes "more or less voluntary systems of behavioral inhibition or restraint." Society is community; society requires other regarding duties in general as its condition. Following Hare, and ultimately Kant, Callicott describes morality in general as "prescriptive or normative" not descriptive.[154] Ethics prescribes limits on behavior. How then can it be derived from ecological studies, based on fact? Callicott himself notes that Hume was the originator of the commonplace notion that an "ought" or norm cannot be derived from an "is" or fact.[155] Callicott regards Hume's solution as acceptable—namely, that ethics are grounded in feeling, not reason, and that this also avoids Moore's "naturalistic fallacy."[156] Moral sentiments are natural, thus universal; as universal, such a basis for ethics is not relativistic. Whether this solves the problems of relativism, the naturalistic fallacy and the is-ought distinction is another question. Moral sentiments are a fact of psychology, as I previously argued, thus an "is," judged relativistic by many critics, and natural by Callicott's own argument for universality.[157] Callicott does remark that the "biologization" of ethics in sociobiology places ethics upon a naturalistic, not relativistic basis.[158] He also argues that nature as "red in tooth and claw" is not amoral in the moral sentiments view. Rather, nature in this sense fits in with the value-free model of science that is without or free of human bias.

As ethics consists of limiting duties, duties to the land can easily be reconciled with traditional ethics. "The acknowledgment of a holistic environmental ethic does not entail that we abrogate our familiar moral obligations to family members, to fellow citizens, to all mankind nor ... domestic animals."[159] Our traditional duties remain. They are expanded to include as morally considerable the nonhuman balance of the biosphere. Two objections can be raised to this answer: that animals cannot reciprocate with mutual duties and that individual rights may conflict with the hierarchy of selection in the land ethic. The latter point was discussed earlier, but conflicts with the rights of specifically human beings remain. On the former point, Callicott argues that duties should be limited to those who can assume them. Duties to infant humans are nonreciprocal; adults as part of a larger community undertake them. The ecology is a community by analogy with the human community. However, recognition is not a necessary condition of reciprocal obligation. Thus even if animals are not recognized by some as part of the community, this does not diminish the obligation, just as in the marginal cases arguments. Against the individualism of the interests view he argues for a notion of "common interests." "All share the common interest of life itself, the desire to live and to be left alone."[160] At a minimum, the land ethic involves letting alone, although other passages suggest some management or interference for the sake of the whole is not out of order (e.g., culling herds).

With respect to human rights, Callicott argues for the idea of different moral communities: human, domestic, and wild. The historically ideal development of ethics has expanded the morally considerable community, but not diminished or eliminated obligations to the smaller communities of which everyone is a part—for example, families and nations. An obligation to feed a domestic animal does not extend to the wild, and may even be in conflict with natural selection. Thus obligations may decrease in number, kind, or extent as the different communities of which one is a part are considered. Humans are part of "mixed communities" in which different relations are involved and to which different obligations are owed. Conflicts over such obligations are, in the last analysis, resolved by consensus. He argues for a "consensus of feeling" among morally responsible individuals based on their sentiments: in effect a majoritarian, not individualistic view. "Radically eccentric value judgments may be said to be abnormal or even incorrect."[161] While this may not preserve all the individual rights argued for by the animal liberationists, conativists, and other value nominalists, it does represent a carefully thought through attempt to preserve the gains of human ethics while undertaking to expand moral considerability to the wild and at the same time respecting the wild for what it is. Further, it represents an alternative to Regan's approach, which not only relocates intrinsic value, but also questions the premise that moral considerability and obligation are coextensive. While the wild is morally considerable, obligation is graduated. Not everything that is morally considerable is due the same or even any obligation (individuals of common species in the wild).

Unfortunately, this argument for degrees of obligation conflicts with another argument of Callicott's, the argument for "moral monism" or consistency, to which we can now turn. Callicott argues against eclecticism, which he calls "moral pluralism,"[162] and for a single moral principle. "Pluralism" is the proposal to recognize different fundamental approaches, for example, animal liberation, ecocentrism, deep ecology, and so on, including one ethic for the wild and another for citizens; one for intimates and another for future generations. These should be grounded in different metaphysics of morals.[163] Callicott understands the motivation of such an approach, and even follows it on an informal level.[164] However, a "univocal" moral principle is required to resolve conflicts, for example, those between individual humans and rare species. Moral pluralism may offer conflicting norms resulting in inaction or contrary actions. Worse, it may tempt toward self-serving actions. "Intuition," rejected as a guide to intrinsic value, is rejected as a way to resolve conflicts as well. Callicott argues that moral pluralism "severs ethical theory from moral philosophy, from the metaphysical foundations in which ethical theory is, whether we are conscious of it or not, grounded."[165] Morals are based on metaphysics, in Callicott's view, and consistency rules out eclecticism. Pluralism is inconsistent because it offers conflicting principles; incoherent, since it is not firmly grounded in science or nature. He urges a consistent principle rule our environmental ethics:

> A univocal ethical theory embedded in a coherent world-view that provides, nevertheless, for a multiplicity of hierarchically ordered and variously 'textured' moral relationships (and thus duties, responsibilities and so on) each corresponding to and supporting multiple, varied and hierarchically ordered social relationships.[166]

Whether his own theory of degrees of obligation based on closeness and intimacy of the relationship meets this test is problematic. In effect, he has argued for one set of obligations for humans, another for domesticated animals, and still another for wild animals. He might urge that these degrees are united by the priority of the land ethic, but it is arguable whether moral sentiments would ever allow humans to be sacrificed for the land. Thus the whole as the source of value or highest good would be sacrificed at a crucial point.

Another conflict is between the "land ethic" and the goal of species preservation. Does the land ethic really prescribe species preservation? What of species that play little or no important role in the various interrelations, but are just a distinct species?[167] How does the land ethic justify their preservation? Isn't species preservation a distinct goal from those consistent with the land ethic—that is, multiple imperatives? A principle of selection between such goals does not *justify* both goals. The functionalism of the land ethic—the role a species plays in the whole—is not necessarily consistent with preservation of non-functional species. Functionalism is an instrumental view, that is, the function a species plays as an instrument in the whole. This may be directly contrary to the intrinsic value of the species that Callicott argues is requisite for preservation. Callicott might point out that the function of a species may be essential, and thus a basis for recognition of its intrinsic value as a species.

As for Callicott's grounding of morals in human sentiments, isn't this just as speciesist and anthropocentric an approach as any? Callicott has argued that anthropocentric ethics is based on certain stipulated psychological states such as pleasure, sentience, or desire. But moral sentiments are themselves psychological states. Further, if intrinsic value is projected by humans, couldn't it be "unprojected" by the same humans if inconvenient? More, is a feeling relation an adequate theory of intrinsic value? How can value be intrinsic to the nonhuman if based on humans and their private feelings? Isn't intrinsic value precisely independent of such feelings, based on the character, essence, or nature of what is valued rather than the valuer? Callicott might argue that although they are based on moral sentiments, they are ascribed to the object, so become independent of the subject. They are a consequence as in the relation of ground and consequent. However, it is difficult to see why such a logical or causal relation escapes the "ontological" problems of a relational theory. If the value is intrinsic to the object, it is not relational, except perhaps the formal axiological relation of intrinsic and instrumental values. Intrinsic value must be independent of any relation to an external entity or it is not intrinsic.

There is also a subtle equivocation in Callicott's phrasing of intrinsic value. This is in the move from projection to the relational aspect of intrinsic value (for itself). To have intrinsic value is to have it projected for Callicott. But then it is not valuable for itself, but for another, the valuer. To jump from projected intrinsic value to 'for itself' is to substitute one relation at a crucial point in the argument, the relation of an instrument to an intrinsic value, for another, the essentially epistemological relation to a subject. It could be argued that this relation to a subject defines the instrumental relation for subjective intrinsic value theories. But then it cannot establish or justify the intrinsic value of a 'for itself.' The subject-object relation is not identical with the instrument-intrinsic relation, although they are equated for some subjective intrinsic value theories. It seems as if Callicott is confusing the two relations. Callicott's example of a baby is a good one, but it could be argued that the baby is a potential valuer and thus accorded legal status based on some sort of conventional consensus. The practice of infanticide in some cultures argues against this as a universal conception of intrinsically valuable.

Another consideration is whether feeling in the form of moral sentiments can be used to ground intrinsic value. Apart from why feelings and only feelings are ultimately valuable, which is borrowed from Hume, there is the question of relativism. Feelings differ from person to person; they are relative to persons and circumstances since the same person might have different feelings about the same object at different times. Callicott argues for a "consensus of feeling" within and between individuals and even cultures. As evidence he uses the widespread injunction against murder in differing cultures, at least in the in-group. Despite variations, there is a "certain normal affective profile"[168] that could be used as a norm. The question remains whether such norms, if they could be demonstrated, could be extended to the nonnatural. The call for an ecological ethic is a good indication that this would be problematic. There would be little need for such an ethic if a consensus on the nonnatural already was in place or could be easily put into practice. No such consensus of feeling about the intrinsic value of the nonhuman exists.[169]

Still another issue is whether the intrinsic value of the nonhuman can be used as a ground for justifying moral obligation. Callicott tacitly borrows utilitarian consequentialist arguments, that the right is based on the good. However, his theory is not utilitarian, although it is close. The minor premise, that intrinsic value entails moral considerability, is not explicitly justified. But it is crucial if obligations to nonhuman elements, however defined, are projected as requiring duties from humans. Indeed, Callicott's graduated scale of obligations argues against just such identification: not all entities with intrinsic value demand duties, at least in the same respect.

In summary, Callicott presents a radically new environmental ethic, which differs in almost all respects from Regan's. The land ethic that Callicott borrows from Leopold is based on ecology, not an extension of older ethical theories such as utilitarianism. Intrinsic value is based on altruistic moral sentiments and projected onto nonhuman bearers. The locus of value is in wholes and species, the biotic community, and includes nonliving elements such as soil, air, and water. Intrinsic value is extended to individuals, but rights are confined to members of human communities, which may include domesticated animals. The duty derived from this approach is to not degrade the land taken as a whole; a principle of selection between elements of the land is clearly stated and unambiguous. A graduated scale of moral obligations based on degree of intimacy produces a more nuanced ethic of the relations of humans to animals.

Certain elements of the "land ethic" persist throughout Callicott's writings. In others, as I have noted, there has been an evolution in his thinking over time. The elements of continuity, which have not changed in different writings, include the need for an environmental ethic, based on a non-anthropocentric theory of intrinsic value. Intrinsic value belongs to nonhuman nature but is projected from a basis in human moral sentiments. The ethic must be modeled on the land ethic of Leopold, and provide a consistent principle of selection in the case of conflict. This must include an obligation to protect the ecosystem as a whole, the "biotic community," as well as endangered species and thus a larger notion of moral considerability. A central duty is not to degrade the land. While wild nature is morally considerable, there is no moral obligation to every individual in the wild, as this would be inconsistent with an ecologically oriented ethic.

Callicott has changed his mind with regard to other issues, namely, the intrinsic value of individuals and the rights of individuals. While in his earliest essays he argues that the value of individuals is in relation to the whole and thus not unqualified, this stance was soon abandoned. In all his later work, from at least 1984 onward, the value of individuals apart from wholes is recognized. Further, rights within the human community may be valid but without validity in the wild—there are different communities with graduated moral considerability. Finally, he argues throughout for a Humean model of the metaphysics of subject and object, but also questions this model.

The biggest weakness in Callicott's approach is in his theory of intrinsic value. The next environmental ethicist I will consider is even more radical than Callicott in his theory of intrinsic value. Because I have considered Callicott at some length, in order to contrast him with the animal liberationists, I will adopt a different procedure for Rolston. I will concentrate mainly on the points in which he differs significantly from Callicott and place less stress on points of similarity.

Rolston: Capturing Natural Value

Holmes Rolston III has also, like Regan and Callicott, called for an environmental ethic. This ethic is to be based on the notion of intrinsic value, an axiological ethic. The reasoning from intrinsic value to moral considerability to obligation is retained. Like Callicott, he has questioned the limitation of intrinsic value and moral standing to higher animals (i.e., Regan's animal liberation ethic), as well as equality of moral consideration. However, he has gone beyond Callicott in three important respects. First, intrinsic value is not based on subjective valuers for Rolston: it is objective in some sense. Second, the locus of intrinsic value lies both in individuals and in larger formations, including species and ecosystems. Finally, Rolston has questioned the link between intrinsic value and moral considerability. Intrinsic value does not confer automatic moral standing on individuals in an ecocentric ethic. This stance makes Rolston the most radical of the three main figures I have considered in scope of value, bearers of value, the ontology of value and the ecocentric standard of moral considerability. Strangely, his position in value theory is close in many respects to that of John Dewey, as he argues for the close relation of intrinsic and instrumental values and how the former often become the latter in ecosystems. Thus although he argues for intrinsic value, and for that matter the objectivity of value, he has a more fluid concept of intrinsic value than subjectively based theories, and ironically is closer to Dewey's instrumentalism, which downplays the importance of intrinsic value.

Many of the reasons for which Rolston calls for an environmental ethic are not peculiar to him. These include the rationales that there is not an adequate ethic for the earth based on traditional anthropocentric ethics, that the latter treats the environment as an instrument, and that an ecological ethic must follow nature, not human practices.[170] The project of deriving duty from value is not new, although his derivation has some interesting twists. Nor is his call for respect for the environment and other duties toward it.[171] His argument that human superiority to other species is not morally relevant is matched by Regan. What is novel is his view that environmental ethics is not peripheral, because human culture is within nature and dependent on it. Nature is the condition of culture,[172] but, more, it is inseparable from culture. Culture and nature are not distinct realms. Thus although nature can be treated like a resource, that is not its final value. Environmental quality is needed for human life, so an environmental ethic is needed. This cuts the ground from under the instrumental or utility view because the environment must be treated more reciprocally as a condition of culture and human life. Rolston has moved from utilitarian to transcendental arguments. In doing so he explicitly mixes is and ought.

How does Rolston differ from Callicott? Why does Rolston go beyond Callicott's fairly radical theory? Rolston's expansion of ethics to the environment involves a new framework for ethics: there is a slight shift in ethical focus that has radical repercussions for the place of humans in the world. Rolston argues, like Callicott, that cooperation among humans is not the model for nature, which involves struggle. However, Leopold's biotic community also includes the human community. Thus a new logic is needed that will include both communities, "a logic and ethic for Earth with its family of life."[173] Ethics traditionally included the individual within a community or web of social relations. Rolston's argument is that this cultural milieu is in turn within nature, and thus that the human community is within the larger natural community. "Culture remains tethered to the biosystem and . . . the options within built environments . . . provide no release from nature."[174] Nature, as the sum total of all physical, chemical, and biological processes includes not only culture, but also human agency,[175] that is, making or technology. Rolston even argues that because human valuing is within ecosystems, it is in some sense part of such ecosystems, and dependent on them.[176] Culture is likened to a "niche" within the ecosystem as it is not and cannot be totally separated from the latter. In effect, Rolston abandons Callicott's basically Cartesian standpoint[177] by attempting to erase the model of a subject in some sense outside of mechanical nature. Humans, their activities, and their culture occupy a special ecological niche, but are not outside of nature. Although ecological ethics is concerned with what Rolston calls "wild values," it impacts culture as both the "frontier" and "foundation" of all ethics as culture and other human endeavor is within nature. The expansion of intrinsic value and moral standing involved in ecocentrism include a shift in perspective away from Cartesian modernism to a view of humans as within a larger whole.

Although Rolston would agree with those authors who argue that humans have greater capacities for a wider range of values, represented in culture, and also are singular in considering the rights of other species, he rejects anthropocentric, biocentric, and egalitarian[178] views of value. Further, he argues that Callicott's view of value is "anthropogenic," in placing its basis in human feeling. This is not a radical enough break with value nominalism and subjectivism. Rolston instead proposes an "autonomous" intrinsic value theory, in which value is independent of human interest or even recognition. Callicott's view places value "extrinsic" to humans but not really intrinsic to the nonhuman. Rolston argues that for Callicott "there is no actual value ownership autonomous to the valued and valuable flower.... A thoroughgoing value theory in environmental ethics is more radical than this; it fully values the objective roots of value with or without their fruits in subjectivity."[179]

Rolston's thesis that value is objective involves the corollary that "nature is an objective value carrier."[180] As the nonhuman carries value autonomously from human valuing, and as much of the nonhuman consists in wild nature, nature must carry or bear values. Indeed, human life and flourishing within the distinctly human ecological niche of culture is evidence for such value to humans. Human values emerged out of the environment in some sense and are still within it in many respects. Such values are a natural event. Rolston argues that from such objective values he can derive duties "binding on all moral agents." An ought is derived from an is: duty from nature with value as the intermediate term. As with many naturalistic value theories, an axiological ethic is the basis for duties but values are based on facts of nature. Rolston's ecological ethic is naturalistic, but with an ecological twist: nature is considered as a whole, not selectively.

The specialized niche of culture has its own ethos. Is this reconcilable with the "wild values" of the biosphere? In what sense are humans within nature? Rolston suggests that human impact can be good or bad, positive or negative.[181] Human impact does not necessarily parallel such natural processes as predation since humans are much more capable of wiping out prey. In a sense all human actions that interfere with natural processes are "unnatural," but some may mimic natural processes (e.g., hunting or childbirth). Even artificial processes can follow nature in a "relative" sense, depending on the "degree of alteration of our environment"[182] and the extent to which nature is present in artificial environments, for example, in landscapes. The mimicking of natural processes is a following of nature: nature as a source of norms.[183]

Rolston also implicitly questions Callicott's ideal of a value-free science. Just as nature may be a source of norms, fact and value are "commingled" in nature. A jaguar in its proper niche may be a norm of a good jaguar: a good of its kind.[184] More, "the mingling of culture and nature means that ethics cannot be isolated from either metaphysics or science."[185] Ethics as a product of culture has not transcended the physical world but lies within it. Natural values can be the object of science, as a form of knowledge that, like science, involves active knowing.[186] Values are in part a form of knowing in which properties of the object are registered. Rolston argues that it is fallacious to "ascribe to the viewer what is really in the scene,"[187] for example, in ascribing beauty to the Grand Canyon. Our valuing is within the environment.

Rolston also breaks with Regan on many points: like Callicott, Rolston prefers an ecological to an animal liberation ethic.[188] Value is not extended *from* humans but encompasses humans within and based on their natural environment. Individuals are neither the sole locus of value nor of moral considerability. "The locus of intrinsic value—the value that is really defended over generations—seems as much in the form of life, the species, as in the individuals, since the individuals are genetically impelled to sacrifice themselves in the interests of reproducing their kind."[189] The rights view is criticized, as in Callicott; the good of ecosystems as a whole are upheld over those of individuals in any conflict. Rolston regards this as a case of deriving an ought from an is, an imperative from the facts of natural relations, for example, between predator and prey.[190]

Human valuing is within nature and a product of it; value also occurs in the nonhuman world. What is natural value? What is its relation to intrinsic value? How can we know such values? Rolston distinguishes between artificial, subjective, natural, objective, and wild values.[191] Artificial value is value added by labor or technique to the naturally occurring value of, for example, wood or gems. Subjective value is emergent from objective value of nature in higher life forms. Wild value is natural objective value that has not been transformed by artifice, for example, in the beauty of scenery. The most significant bearer or "carrier" of objective value is nature: "natural value is ecological-relational."[192] This might seem like a relational theory as in Callicott, but Rolston explicitly rejects such a view. The relational qualification involves interrelations in the environment—for example, between species, niches, and the like. Natural relations receive as much stress as individuals and species: the ecosystem as a whole. Rolston mentions several criteria of such value including a "project of quality" of a species or individual and the "intense harmony" between species that marks them as "good in community."[193] An individual or a species may have its own good and may also play a valuable role in the ecology in relation to other species. Natural good is multiform, but the various types are interrelated. The food chain involves such relations: lower life forms "sacrifice value" to higher predators. Such value relations are independent of human valuing, although human valuing is dependent on them, thus they are "objectively present in ecosystems."[194]

Sacrifice of value is instrumental, but Rolston argues that intrinsic value is also present in the project of the individual of a species. This value is primarily connected with life. "A life is defended intrinsically . . . [in] a form of life as an end in itself."[195] Rolston's stress on forms of life, that is, on species, allows him to view individuals as containing the intrinsic value of the species, while allowing for the death of individuals, including the capture of its intrinsic value by predators. The individual in nature is a "good of its kind;" its project is the project of the species instantiated in an individual. The defense of an individual life is the defense of intrinsic value; death is the capture of such value. An organism has "value ownership of its own life," such that "each organism has its own good."[196] This represents the specific nature "programmed" into the individual such that the individual "exemplifies" the species. Organisms have their "own standards" and "promote their own realization" as a "good of its kind; it defends its own kind as a good kind."[197] Defense of its own life is a defense of a good of a kind, a specific good. The life is intrinsic to the individual organism, but related to the species. Rolston argues that "everything with a good-of-its-kind is a good kind and thereby has value."[198] Even invasive organisms have a good of their kind. Life consists of organisms embodying species that are "spontaneous evaluative systems" and "genuinely autonomous."[199] To value wild life is not subjective; Rolston argues that value should be extended from "subjective people experiences" to "objective . . . lives."[200] Value includes both objective natural value and subjective natural value.

What of nonliving nature? Although Rolston sees value in the creativity of nature,[201] in its diversity, beauty, and resourcefulness, intrinsic value does not seem to attach to nonliving nature. A fossil has only instrumental value.[202] Though vital to life, soil, air, and the like are instruments. This view is less radical than Callicott's and brings Rolston closer to the biocentric vitalists and conativists. However, Rolston does not share the value nominalism of many of the latter.

Higher than the good-of-its-kind of a specific individual is systemic good—the good of the whole.[203] This is the intrinsic value of the "value producing evolutionary ecosystem, the historical environment in which humans were generated."[204] "Nature is valuable intrinsically as a projective system. . . . The system is of value for its capacity to . . . pro-ject all the storied natural history," that is, the "inventiveness of systemic nature . . . at the root of all value."[205] The value of the whole, at least the whole of life, is greater than and prior to the "parts" or specific roles, niches, and relationships in the ecosystem. The objective systemic process is an overriding value not because it is indifferent to individuals, but because "the process is both prior to and productive of individuality."[206]

Objectivity of value means that Rolston must question the premise of modern value theory: that value requires an evaluator. The further premise that value requires some sort of consciousness or awareness is also questioned; Rolston argues for value in "an informed process in which centered individuality or sentience is absent."[207] He compares this holistic understanding to economic markets that function without conscious direction or the structure of individuality, but measure value. Although there are no rights before the advent of humans, "values (interests, desires, needs satisfied, welfare at stake) may be there apart from human presence."[208] This quote gives some insight into what Rolston means by value, including many of the standard marks of subjective, naturalistic value, for example, interests, desires, and needs. These are interpreted objectively, however, as belonging to nonhuman individuals, for example, the desire of an animal for food or the need of a plant for water to flourish. There is an implicit denial of the subjectivity of such natural states. They are not exclusively human: life support and genetic information operate regardless of human valuing and awareness.[209] Rolston is not claiming that humans cannot know natural value, only that natural value is independent of humans. "Insentient organisms are the holders of value although not the beholders of value."[210] Description of an ecosystem and evaluation "arise together": ecological description informs ecological values. Value is not imposed by humans but discovered with the discovery of natural kinds and relations. The normative is not so much derived from nature—ought from is—as discovered with it. Or, more exactly, the process moves from the 'is good' to the ought, the moral.[211] Ecological or natural value is firmly grounded in the object.

Value is a quality, but has a range from subjective to objective qualities. Some values seem more "objective" (e.g., nutritional value), while others are less so. Value is "specific" rather than generic: a specific, adjectival qualification of an object.[212] Such qualities "inhere" in value holders.[213] Whether this qualitative view is compatible with the relational view involved in ecological niches and roles, a point that was raised with respect to Callicott, is another question. I will return to this point. Another point is that a range of qualities can exhibit value, which Rolston outlines in some detail. These include values ranging from "individual preference value," which is similar to subjective value in the most reductive sense to "ecosystemic value," which is neither individualistic nor subjective.[214] The question arises of whether a quality in a subjective or objective individual and ecosystemic whole is a quality in the same sense. However, Rolston has a historical and emergent view of natural history and thus of value, so it is plausible to argue that different qualities emerge at different times and that what is constant is the whole, the system of interrelations. Linguistically all are qualities, as they can be predicated of a subject in the grammatical sense of the latter.[215]

Rolston covers all the aspects involved in the relation of instrumental to intrinsic value; he does not differ significantly from Regan or Callicott in this respect. That is, the relation of instrumental to intrinsic value seems to be the same as the relation of means to ends (the teleological relation) and that of for (the sake of) something else and for itself (for its own sake), the reflexive relation.[216] Value is attached to the "thing in itself," "for what it is in itself."[217] In the case of objective and holistic qualities, the "thing itself" is something living: "a life defended for what it is in itself, without further . . . reference." Organisms have "value ownership" of their own life.[218]

Rolston does not break new ground on the ontology of value either, if both Regan and Callicott are the standard of reference. That is, objective value is "in" nonhuman life,[219] objective standards are "in" the organism.[220] Like Callicott, Rolston states that nonhuman life and organic systems "are" (of) value, the copula seemingly forming an objective proposition.[221] Like Regan, he also uses the language of having: "trees have their norms. . . ."[222] Value may be added as a quality, both artificially, as in economic value and the value added to natural value,[223] and also as an existing "excellence" of nature.[224] Combining Regan's and Callicott's approaches, Rolston argues for the value of individuals, species, and larger wholes such as ecosystems. "The living individual, taken as 'point experience' in the web of interconnected life, is *per se* an intrinsic value."[225] Like Regan he attaches this value to the life of the individual. But Rolston argues just as strongly for the value of the species or kind[226] and of the systemic whole.

Like Callicott and unlike Regan, Rolston argues for the value of the whole or "system" over the species and the species over the individual. The individual "instances" higher values of the species,[227] and is more easily replaced than species with their niches, to speak nothing of the ecosystem as a whole. The justification of this hierarchy of degrees of intrinsic value is that the whole is greater than the sum of its parts. "The system is a web where loci of intrinsic value are meshed in a network of instrumental value."[228] In other words, intrinsically valuable individuals become instruments from a systemic point of view. The system is "of value" apart from individuals and species within it as it is "productive" of the distinctive values represented by species and individuals. This model of the relation of intrinsic to instrumental values comes close to Dewey's critique of intrinsic value as ultimately a form of instrumental value, a topic I will cover in chapter three.

What is new with Rolston is the view of culture as within nature; thus subjective value is within natural value and an emergent property of natural ecosystems.[229] The ecosystem holds values 'within' itself as a "carrier," "producer" and "holder" of value. Nature "carries" values both apart from human valuing and as value for human use.[230] But the values carried by nature are not subjective; the earth carried value prior to humans. Carry here seems to imply that nonhuman living things 'have' value within an ecosystemic whole. The earth or the whole carries such value loci within it. Thus organic and ecosystemic value is basic to all other kinds,[231] as the condition of them. The earth has positive value as a "satisfactory" environment for humans, even if culture creates a complicated subhabitat. Rolston gives a long list of values carried by nature, such as "life support value," "recreational value," "aesthetic value," and so on.[232]

The ecosystem is also "productive" of value; production is not confined to human making but also includes the production of new species, roles, niches, habitats, and so on, all of which create the web of life from which human valuing emerged.[233] "In an ecological perspective, that Earth is valuable would mean that Earth is able to produce value and has long been doing so as an evolutionary ecosystem. . . . When humans come they find Earth often valuable, able to satisfy preferences, able to produce valued experiences."[234] This may imply an instrumental relation, but the value of the ecosystem apart from human valuing has already been argued. Humans are a late addition to this productive system and emerged within it. The argument is that if they are of value, the process that produced them, which produced value, must be *a fortiori* of value. Its value is in its production of different kinds of value and as the basic condition of such value. Further, other kinds are modifications of natural value, for example, economic.[235] The fact of modification may also be considered a natural value, that technique is possible.

Rolston outlines different "zones" and levels of value. The hierarchy of values is not simply axiological, embodying degrees of value, but also of degrees of being[236] and of internality or subjectivity. That is, there is an ascending scale of values from the lowest life form to the highest states of consciousness in which subjectivity of value is a mark of a higher form of being and value. The modern tradition of subjectivity is incorporated rather than reduced: it is emergent from lower forms of intrinsic value in the natural process of evolution,[237] and, while a mark of higher value, it is not exclusively of intrinsic value. He avoids the problem of a conflict between humans and wild animals in that the latter are granted intrinsic value, but not as much as humans. There is a degree of instrumentality to "lower" beings that decreases as one encounters higher and higher organisms, while intrinsic value increases by the same proportion.[238] Further, there is value in the organic, in the uniqueness of life as such,[239] not just in levels of life or relative intrinsic value in species. There is a "valuational complexity" in the whole with different degrees, zones, and dimensions. Zones refer to the degree of penetration of the wild into culture, for example, urban, rural, and wild,[240] while dimensions refer to the complex of value bearers, zones, degrees, and kinds. Thus an individual subjective value is one dimension while the value of a unique habitat is another.[241]

One potential problem here is the conflict between human values and natural values, especially between what is highest ecologically, the ecosystem, and in degree, that is, human subjective valuing. Which is higher? Rolston insists that the complexity of value dimensions be kept in mind in any conflict.[242]

Although humans have priority over other species on the whole as having personality and realizing a greater range of values, subjective value is not the only value. The value of the whole may be more important than certain subjective values, for example, economic demand values.[243] Rolston's incorporation of levels and degrees of values may let in through the back door what he was initially trying to keep out—an instrumental view of nature. The arguments for the greater intrinsic value of higher beings and personalities are, respectively, Aristotle's and Kant's arguments for the instrumentality of animals and other lower life forms.[244] Rolston is arguing for a more graduated view of the transition from instrumental to intrinsic value, but such a realist hierarchy lends support to an instrumental view. This is especially the case if intrinsic value slides into instrumental value at lower levels, as it clearly does in Rolston. Nor is a condition of a value necessarily either of value or of intrinsic value.[245] Of course objective value can attach to instruments, as Dewey and Lewis argue; intrinsic value and objective value are distinct. But Rolston is arguing for objective intrinsic value as well.

The complexity of value dimensions applies to the locus of value as well. It is evident from his complicated theory of value, based on the web of relations in the wild and their relation to culture, that the locus of value will be at different levels. There are loci of value "across the whole continuum," from individual to the ecosystem as a whole and from the instrumentality of non-living nature to the subjectivity of culture.[246] An individual is both intrinsically valuable and a potential instrument of a higher form of life. In itself it is a locus of value as an individual, a "concrete attempt at problem solving," with a unique genetic makeup in unique circumstances at a particular point in history.[247] It is also a token of the species, that is, the locus of species value.[248] If its value is "captured" in predation and used instrumentally, it becomes a locus of ecosystemic value, "used instrumentally to produce still higher intrinsic values." In this sense, "value is what is favorable to an ecosystem, enriching it, making it more beautiful, diverse, harmonious, intricate."[249] Thus the direction of value is upward both in the emergence of higher dimensions of value, such as subjective states, in evolution, and the value of instrumentality in contributing to the value-producing system of life as a whole. Because he is basing natural value on actual ecosystemic relations, he can use the fluidity and dynamic of the relations within it between individuals in separate niches to move easily between instrumental and intrinsic value. Like Dewey, he does not abandon the notion of intrinsic value, but modifies it, so that the contrast between intrinsic and instrumental is not absolute or fixed. Unlike Regan, he is not radical on the insularity of intrinsic value.

Rolston speaks of all sorts of "intrinsic values" and their location in very different levels, although the token or bearer of these levels may often be the same individual. He notes that although the individual having intrinsic value is "in some sense on its own ... [it] is in another sense embedded in an environment."[250] Besides embodiment in individuals by themselves and as bearers of specific intrinsic value, there is a distinctive "organic" or systemic intrinsic value, which I discussed earlier. This consists in a holistic matrix or web of relations such as the food chain. Rolston notes that the intrinsic value of the individual is "problematic in a holistic web."[251] The system as a whole contains the other values which have a dynamic relation within it. Rolston describes how within the system parts are used instrumentally by organisms that "value something intrinsically; their selves, their form of life."[252] The instrumental-intrinsic relation is natural, embedded in the ecosystem in the form of "value capture and transformation," characterized by the "competing, exchanging and intermeshing of goods." The bearer of ecosystemic intrinsic value is not individuals as such but individuals in relation. Their value is captured and serves higher values: "intrinsic and instrumental values shuttle back and forth, parts in wholes and wholes in parts."[253] Intrinsic value is located not in individuals simple but in them and also in their relations within the matrix of life. Intrinsic value can become instrumental and in a sense is embodied in the same individual. Intrinsic value is still in individuals where a life is defended, but this value is relative to a larger whole.

The justification of this complicated notion of dynamic locus is that the wild is not culture and has its own norms. Indeed, Rolston can describe wild nature in terms of "amoral normative systems" by comparison with the individual rights of culture. Nevertheless, value is produced, conserved, and advanced. The process "throws up" emergent qualities, such as human subjective value as well as "goods of their kind." Dynamic intrinsic value produces higher value. Species benefit in the long run even if individuals are sacrificed. The system is a "value-shunting device," in which value is captured for the good of the whole: value "transformation and circulation."[254] Over time value is produced over and above individuals even though they are temporary bearers of intrinsic value. The "indifference" of nature to the individual is part of the overall pro-ject. Intrinsic value is genuine, but is located ultimately in "projects of quality" that involve "creativity," whatever their locus, individual, relational or holistic, subjective or objective. "There is value wherever there is positive creativity."[255]

Such a wide and dynamic notion of the kinds, bearers, and locus of value enable Rolston to criticize a wide array of traditional value theories. These are centered on subjective,[256] anthropocentric,[257] nominalistic,[258] and individual rights theories. However, Rolston also rejects Moore's realism[259] and the idea that value is emergent.[260] Subjective value is emergent, but not natural and wild value. Rolston's criticism of Animal Liberation and its extension of rights to animals parallels Callicott's.[261]

Rolston's theory of dynamic intrinsic value attenuates the connection of intrinsic value to moral considerability. The disconnection is based in part on his dynamic-natural notion of intrinsic value in which intrinsic value can become instrumental. Rolston attempts to justify value from an ecocentric point of view, as the matrix of interconnections, which form an ecological system, is not centered and thus diffuses value away from an exclusive focus on individuals.[262] Further, nature operates on a different ethos than civil society, even if it can be a value object and even if the latter involves exploitation of the former based on its principles, that is, amoral predation and exploitation. The "indifference" of nature to struggle and death is not a suitable ethic.[263] Rather than deny individual intrinsic value, the early solution of Callicott's, Rolston wants to retain intrinsic value of individuals in a weakened form. Because this is a relatively weak notion of intrinsic value, moral considerability is also weakened except at the upper end of the hierarchy of values. That is, individuals have intrinsic value, but, at least in the wild, this does not give them moral standing, despite certain statements to the contrary.[264]

Further, Rolston tacitly breaks with the long tradition in ethics, encompassing a wide variety of ethical theories, which posits a necessary relation between intrinsic value and moral considerability, although it may only be a qualification of it from a naturalistic perspective. Rolston argues that individuals with intrinsic value may nevertheless be sacrificed for the good of the whole: their intrinsic value either does not entitle them to moral considerability or is of subordinate value even if intrinsic to higher intrinsic values. The first thesis is the more radical since if the connection of intrinsic value to moral standing is lost, almost the entire tradition of Western ethics is undermined.[265] Rolston might argue that he is only qualifying the theory, not undercutting it, because he is only arguing for a hierarchy of value within intrinsic value, a seeming commonplace, for example, in the "highest good" thesis. However, intrinsic value is equally distributed in most such theories[266] while subordinate goods are usually deemed instrumental (e.g., in the relation of animals as means for humans in Aristotle and Kant). Rolston's insistence on claiming intrinsic value for individuals while subordinating them to higher goods like the ecosystem threatens to undermine intrinsic value of the nonhuman along Deweyan lines or their moral considerability. Such an outcome would defeat his initial project. Regan and Callicott more consistently view intrinsic value as either individualistic or holistic, respectively. This radical break in the connection between intrinsic value and moral standing is perhaps the least satisfactory part of Rolston's theory since it tends to undermine a relation that, by consensus, is required for duties to the nonhuman. However, it does resolve the relation between the wild and civil zones, in that neither is imposed on the other;[267] different dimensions entail different responsibilities. The apical standing of humans in terms of evolution and also the hierarchy of values do not mean that humans escape responsibility for the rest of life.

Rolston notes the difficulty of defining duties to an amorphous, dynamic, and intangible thing such as an organic ecosystem, despite analogies to human communities.[268] He calls for creativity in ethical thinking and continual input of new factors and evaluations. The long list of human duties to the environment that Rolston covers have two basic principles. The principles reflect degrees of obligation based on degree of intrinsic value or a grading of interests. Rolston believes that the interests of nonhuman life can be graded and their "welfare" taken into account by human moral agents.[269] Only the highest beings with intrinsic value have full moral standing and thus are due complete moral obligations. Second, obligations should accord with the character of the dynamic system. Such obligations are "not categorical, not absolute and independent of circumstances and beliefs about the world. They are systemic and the systemic components are both natural and cultural."[270] Natural obligation does not resemble the absolute categorical imperative of Kantian personalism, as intrinsic value is not the mark of an absolute end in itself. But the cultural is involved in the fact of a nonnatural ethic.

The degrees of obligation are difficult to assess with precision, Rolston admits.[271] It requires appraisal of what we do not personally value, that is, objectivity. It requires that we go beyond our preference for our own kind and extend ethics to other species and the web of relations in which we also share. In general it is based on the degree of sentience and subjectivity. As I have noted, it does not extend to individual animals, especially in the wild. Above this, there is a hierarchy of obligation, to species, particularly rare ones, to habitats and the ecosystem, and to other humans. Rolston argues firstly for preservation: "the species-environment complex ought to be preserved because it is the generative context of value."[272] Preservation of species is called for as they are of greater uniqueness than individuals; preservation of habitats is essential to preservation of species. Duties to the ecosystem differ from duties to humans as the welfare of the ecosystem may involve sacrifice of individuals, a point that echoes Callicott. However, wild value is greater in proportion to its rarity and the threat of irreversible change to the ecosystem as a whole.[273] At the apex of moral obligation stands the "projective system" as the most valuable of all, although human subjects are its most valuable "products."[274]

It was argued earlier that this might create conflicts of duties, and Rolston exerts considerable effort to considering such conflicts of duties and ranking them. The obligation to protect other humans comes before any obligation to individual animals and plants, and the obligation to protect species above individuals and rare species above common ones.[275] Above the individual level, however, duties become complicated. Rolston argues that an absolute obligation to human life would entail unlimited human births at no matter what expense to the ecosystem. This is unacceptable and thus the obligation to the human species, though paramount, is not absolute. There ought to be limits on human desires, interests, and pleasures if their impact on the ecosystem or rare species is catastrophic.[276] Rolston admits that a rare species may have no resource value and its loss may not even affect an ecosystem in the long run. He argues, however, that ethics concerns obligation, not human preferences, and that limiting the latter to save the former is justified as an obligation.[277] Instead of the equal value of all life, equal respect,[278] Rolston summarizes his view as entailing the "principle of the non-loss of goods." Goods embodied in rare species entail moral obligations that may outweigh specific human interests. The good accomplished by destruction must in some sense outweigh the goods in the organism destroyed.[279]

Regan has a fairly thoroughgoing list of specific duties to the environment, including duties to animals,[280] the land,[281] the wild,[282] endangered species,[283] and ecosystems.[284] Various principles of priority in case of conflicts between human interests and the environment are outlined as well as special obligations for commerce.[285] Rolston calls for collective choice in an ethic of the "commons," which is wild land considered as a common resource.[286]

Rolston has not been without his critics, including a seeming ally, Callicott. Callicott has argued[287] that Rolston is a Kantian but that his theory of value is inconsistent with Kant's, that he is unscientific, and that he ignores the fact-value distinction. Finally, Rolston presumes and does not overcome the modern worldview derived from Descartes, yet his ethic is inconsistent with this view. On the first point, whether Rolston is consistent with Kant, Rolston can easily be defended: he is not a Kantian in decisive respects.[288] His theory of value may be conative, as Callicott claims, although Callicott stretches the latter term to include quite distinctive theories, for example, biocentrism and the interest theory. In any case, Kant's theory of value is neither conative nor vitalist. Nor does Kant extend moral considerability to animals, although this is true of almost all traditional theories and Callicott's point may be that Rolston adopts the framework of Kantian ethics, not its theory of moral considerability. Finally, Rolston rejects the "Categorical" absoluteness of obligation, as I pointed out earlier.[289] Although there are transcendental arguments (e.g., the condition of subjectivity is the ecosystem as a whole),[290] the overall argument is consequentialist, not deontological: duties are consequent on value[291] and are formulated in the light of specific outcomes. Rolston might be criticized for inconsistency in sneaking in the form of Kantian arguments, but this is not logically inconsistent.

With regard to Rolston being unscientific, it is clear that he bases his environmental ethic, even more than Callicott, on the ecology. What he tends to ignore is physics and the other hard sciences, which Callicott cannot seem to get around. My own impression is that, on the contrary, Rolston is very well aware of the positivist-empirical view of science, which remains, between the lines, hovering in the background of his thought. This accounts for the premises in subjective value he begins with, the almost apologetic and tentative tone of his extensions of value and moral standing, and his attempt to bring in physical theories at odd points. What he is not aware of is alternative views of science, for example, those of the pragmatists. Callicott's insistence on the value-free model of science has been criticized by various pragmatists as an inaccurate model of actual scientific practice. Rolston's theories are actually incompatible only with the positivist view of science, a point in their favor.

Callicott is himself inconsistent on the fact-value distinction, as was Hume, its inventor. Moral sentiments are facts of human nature or psychology. If value attaches to them, the distinction is invalidated. Rolston in fact only attenuates the distinction, he does not deny it, as I argued earlier.

Finally, Rolston is attempting to overcome the Cartesian worldview. True, he starts from it as the context of his discussion, but his break with modern ethics and its anthropocentric, egoistic, conscious subject, is decisive, and even more radical than Callicott's. The subject and object are retained as points of reference, but the subject is re-placed in the objective world, and is reduced to a difference in degree not kind. The emergence of the subject is within the context of nature: the subject naturalized.

A more basic criticism of Rolston is whether his is an adequate theory of intrinsic value. If intrinsic value easily becomes instrumental value, can he adequately defend his theory from Dewey and others who attenuate the distinction between intrinsic to instrumental value? Can he adequately defend the intrinsic value of objective individuals if it is sacrificed to predators for the good of the whole? In what sense do individuals on the one hand and species and wholes on the other share some value quality? Is intrinsic value a quality of life or is life the criterion of intrinsic value? Just which quality or attribute has intrinsic value—life itself or a living thing? How can duty derive from intrinsic value if intrinsic value can be instrumental? Wouldn't Rolston have done better to argue that the intrinsic value of individuals is the specific, general good, and that wild individuals contain no value qua individual but only as bearers of specific value, a consistent holism? Isn't his argument for ecosystemic value an excellent argument against individual objective natural value? Again, as with Callicott, the arguments for holism may not have the outcome of preservation of rare species. These are two distinct principles, based on distinct value loci, rarity of a type versus its hierarchical place. As for hierarchy, his arguments for it fit neatly in with typically anthropocentric arguments for human predominance and supreme value.

Another group of criticisms can be made with respect to the place of humans in the natural world. Rolston excepts humans from the necessity of sacrificing individuals for the good of the whole, but does not consider examples of humans doing just that: in war, capital punishment, abortion, and infanticide. It may be that these examples are exceptional, but they are *prima facie* evidence that the culture-nature split may not be as sharp as he makes it on this point. Further, humans eat meat based on analogy to natural processes of predation, but this undermines the separateness of culture in another respect. More, should humans follow natural processes or cultural norms? If they follow the former in one respect, why not in all? What is the principle of separation?

In summary, Rolston has offered a modified or secondary[292] axiological environmental ethic in which value is derived from and congruent with natural, ecosystemic processes and duty is derived consequentially from value. Rolston is more radical in his naturalistic theory of value than either Regan or Callicott, both in its scope, living organisms of all types, in subjects and objects; and in its locus in everything from individuals, through species and up to ecosystems. He rejects the limitation of intrinsic value to higher animals in Regan and its source in the subject in Callicott. He is even more radical in his notion of moral considerability, based on the dynamic of intrinsic value within natural processes of the ecosystem. In this position, as I argued, he approaches Dewey's naturalism with its rejection of ethical grounding in intrinsic value. Moral considerability is weakened in Rolston by the inclusion of all living individuals as having intrinsic value because that intrinsic value does not entitle them to moral consideration. However, the locus of moral considerability is similar to Callicott's. Rolston is more detailed and thoroughgoing in his analysis of distinctive duties to the environment. He is Kantian only in his occasional use of transcendental arguments, including the priority of duty over interests and pleasures, and a single appeal to the Kantian value, autonomy.[293] In truth Rolston represents a return to a naturalistic ethic, albeit with a twist. Rolston's naturalism is general not specific. It is not based on specifically human nature, but rather on nature taken as a whole or at least the biosphere as a whole with its collective community of species. It is a nonspeciesist naturalism, breaking with the Aristotelian tradition of species naturalism. In this it may also approach Deweyan naturalism, a point I will examine in a subsequent chapter. Rolston's naturalism is also the most radical in its assessment of the human condition as within the larger environment, in effect a "postmodern" break with modernism and its egoistic, almost solipsistic subject.

Norton and Anthropocentric Environmental Ethics

Bryan Norton has defended what he terms a "weak" anthropocentric environmental ethics and has included a critique of non-anthropocentrism in his major work on environmental ethics.[294] Norton agrees with other environmental ethicists that an environmental ethics is needed, particularly with respect to the issue of species preservation.[295] He does not dispute the idea that there are some things with intrinsic value,[296] that intrinsic value is the basis for ethical obligations[297] and that value concepts are the framework of and relevant to environmental issues.[298] Where he disagrees with the three authors whom I have discussed in this chapter is on the issue of the scope, bearer, and locus of value. Norton's theory is anthropocentric, and he claims that an anthropocentric value theory is sufficient to justify species preservation. Not only are non-anthropocentric theories not required to justify species preservation, they are fallacious. In order to examine these points, I will first briefly describe Norton's theory and then discuss his criticisms of non-anthropocentric theories.

Norton defines anthropocentric in relation to non-anthropocentric. "Value premises and the axiological systems supporting them can be characterized further as 'anthropocentric' or 'nonanthropocentric' depending on whether all value countenanced by them derives ultimately from human values."[299] Humans are not only the basis, bearer, or origin of value but also the sole locus of intrinsic value.[300] Instrumental value derives its value from the relation to human intrinsic value. According to Norton, no theory contests the idea that individual humans have intrinsic value; thus it does not stand in need of justification.[301] The attempt to extend intrinsic value to nonhumans is based on analogy to humans. Like Callicott, Norton argues for a relational theory, in which (instrumental) value may be independent of humans but depends on the human experience of value, that is, a relation of a valued object to a valuer or subject.[302] Norton characterizes this position as "weak" anthropocentrism because some value is accorded nonhuman objects. However, intrinsic value is confined to humans; nature does not have intrinsic value. As I noted, Norton accepts the idea that intrinsic value entails some sort of obligation. Although he argues against consciousness as the criterion of intrinsic value, that is, as the specifically human difference,[303] he still defends a form of subjectivity as the basis of value, that is, as the end or intrinsically valuable. Nonhumans are not subjects as they cannot "consider and reject value commitments."[304] Subjectivity of intrinsic value does not mean that *intrinsic* value is dependent on other subjects; rather, it is located in subjects. Subjects are its bearers and other values are in relation to such intrinsically valuable subjects. "However ethical theorists interpret the intrinsic value of human individuals, they all agree that this value does not depend upon the individuals' attractiveness to other humans."[305] Further, basing value on subjective valuers does not imply that the values of the valuer must be uncritically accepted, a point I will return to. Values may be independent of a particular valuer. He even argues that the value of the object to the valuer may result from certain "qualities it has," a position close to Lewis's and even Callicott's.

In defending subjectivity of intrinsic value, Norton argues that the subjectivity of value is not a ground for discounting it. Just as the subjectivity of beauty does not devalue beauty, so are species required even for subjective valuing of them.[306] He notes that much of the dispute over values between anthropocentrists and non-anthropocentrists is whether an anthropocentric theory is sufficient to justify limits to human exploitation of the nonhuman. He argues that if a rationale for species protection can be derived exclusively from an anthropocentric value theory, from "human ideals," there would be no reason to attribute value to them.[307]

Norton divides instrumental values into demand and transformative values. Demand value is borrowed from economics and refers to satisfaction of "preferences." Transformative value "provides an occasion for examining or altering a felt preference rather than simply satisfying it."[308] Demand values can be evaluated for felt but unconsidered preferences and also deliberative, considered preferences. Transformative values have a role in such deliberation by altering preferences. Not all felt preferences survive evaluation as worthy. Only justified demand values have a claim in policy formation. "A value system that includes transformative as well as demand values can limit and sort demand values according to their legitimacy within a rational world view."[309] "More rational" preferences may override demand; rationality is the principle of evaluation between competing values. According to Norton, nature provides such transformative values, as the experience of nature "elevates human values" and thus is of greater value to humans.[310]

A view that only considers demand values Norton links to "strong" anthropocentrism; including transformative values in a theory is a mark of "weak" anthropocentrism. He uses the demand/transformative classification to characterize arguments for species protection, that is, as arguments based on demand value for humans of species protection and the transformative value for humans of species as justification for species protection. Much of Norton's effort goes into arguing against the sufficiency of demand value and the need to supplement demand with transformative value. Partly this consists in arguments against benefit-cost analysis as an "objective" measure of demand value.[311] It is difficult to measure the costs of extinction even with respect of "willingness to pay"—the criterion of benefit cost analysis.[312] The role of nonhuman nature in satisfying demand is not enough to justify preservation of species, in his view.

Applying this distinction to the debate over preservation of species, Norton argues for the transformative value of the wild. This value in turn requires the belief that some values are "objectively better than others," but not in the intrinsic value of nature.[313] "Objectively better" anthropocentric values are the result of transformation. One thing not measured by demand value is the "existence value" that "attaches to the knowledge that a species exists," without reference to its use.[314] Wild nature has "contributory value" to human development as well as being an instrument for human survival. "Species preservationists can argue that overly consumptive demand values are less rational because their fulfillment threatens the system within which the human species has evolved and must continue to exist."[315] Thus conservation is more rational as humans are still dependent on their environment. By positing transformative values, Norton is upholding a "middle way" between the radical intrinsic value ecologists and the economists who have attempted to reduce all value to demand value and justify human exploitation of it. There are acceptable and unacceptable uses of the environment.[316]

In a separate part of his book, Norton critically considers three distinct theories of the intrinsic value of the nonhuman to determine if such theories are "plausible." He notes the widespread use in such theories of "arguments from marginal cases" and the argument that nonhumans do not differ in relevant respects from humans in the characteristics needed for intrinsic value, for example, sentience.[317] Non-anthropocentric theories often borrow value theories from the tradition and extend intrinsic value to wider scope, that is, to include nonhumans. Norton poses three questions that he subsequently addresses to each of the three main theories he examines: what is the locus of value, "what does it mean to ascribe intrinsic value to an object," and how are such ascriptions justified?[318] It is notable that he uses the term "object," rather than animal, living being, or ecology as that to which intrinsic value is to be ascribed. He also sets two conditions that any such theory must meet: that ascriptions of intrinsic value must be justified under some "plausible theory of intrinsic value"; second, "the type and range of intrinsic value attributed must be sufficient to justify characteristic claims by species preservationists."[319] Norton also provides a chart of possible positions on the loci of value, from individuals alone to individuals, species, and ecosystems. Combined with whether such loci are based on value theories that are analogous to anthropocentric theories, he lists fourteen permutations of locus and theory.[320] Norton considers three authors on intrinsic value whom he argues exemplify his list.

The first author he considers is Peter Singer. Singer shares with Tom Regan the "animal liberation" view. However, Singer is a utilitarian, not an individual rights theorist. Thus some of Norton's criticisms of Singer are not relevant to this chapter.[321] Others, however, are germane, such as the argument, also made by others, such as Callicott, that animal liberation in general does not provide a justification for preservation of rare species that are not "sentient." Thus rare plants would have no moral standing on such a theory. Further, as Callicott has also argued, predation and the culling of herds could not be justified on the basis of nominalistic value locus—that is, individual rights.[322] The argument from benefits to humans also does not support preservation of rare but useless species. In general, Norton finds Singer's theory plausible but not feasible for preserving species.

A second theory is that of Paul Taylor.[323] Although his theory differs in important respects from Holmes Rolston's theory, there are many similarities, including an objective notion of intrinsic value and a wide-ranging scope for the locus of value to include individuals, species, and ecosystems.[324] Like Rolston, he sees organisms as "teleological centers of life" and argues that any attempt to separate humans and animals on this score is an irrational prejudice.[325] Norton criticizes Taylor for not making the case for a necessary relation between having a "good of one's own," the latter's criterion of intrinsic worth, and having intrinsic value. Norton denies there is a necessary relation.[326] He also denies that an individual can be an end in itself in the wild, a claim echoed by Callicott. Without some differentiation of humans and nonhumans, the justification of culling of herds in terms of the value of an ecosystem would justify the culling of excess humans.[127] Finally, a theory based on the value of individuals, even if the value of larger wholes is justified in terms of individual value, cannot provide a rationale for species preservation, as it starts from individually based ethical theories.

Since individualism has proven inadequate, Norton turns to Callicott's theory, which attempts to attribute value to species and ecosystems. Norton has argued for value within the tradition of "interests" and finds it hard to clarify what the notion of the interest of a species would be. He notes that analogies to individual interests are not helpful in a holistic context. The interest of a species would have to be apart from the individuals of which it is composed; thus analogies to corporations are not apropos.[328] If a holistic ethic is articulated separately from individualistic ethics, it may have negative consequences for humans. Further, a holistic ethic, which Norton claims to find in Callicott's theory, is not consistent with his moral sentiments view.[329] The latter is inadequate as a justification of species preservation, especially in "controversial cases." Norton concludes that non-anthropocentric views of intrinsic value are "premature."

Can non-anthropocentric theories of intrinsic value be defended from Norton's criticisms? What immediately strikes one about Norton's analysis is that it assumes a Cartesian framework of subjects and objects. The distinction of subjects and objects is the origin of the fact-value distinction. Yet Norton admits that its separation of fact from value is inadequate as a model for the environment and that ecological studies may inculcate attitudes that transform consciousness of the environment. Further, scientific knowledge serves demand values: it is not purely for information. Thus the relation of ought to is may be more complicated and it may be feasible to argue the objective case of derivation of values from (living) objects. Norton's value framework is utilitarian-economic, for example, in his analysis of value in terms of interests, but by his own arguments the demand value that interests specify is inadequate as either a comprehensive theory of value or a justification of species preservation. The rational qualification to demand values is not consistent with either demand values or transformative values. If rationality is the ultimate standard, demand and transformative values cannot collectively form a theory of value. Nor, for that matter, are demand and transformative values one theory, as one is based on preferences and the other on suppressing or eliminating exactly such preferences.

The criterion of sentience is utilitarian, that is, value is based on capacity to feel pleasure and pain. However, this subjective state is not equivalent to individually based value. Like most subjective theories, utilitarianism seems to posit a connection between the intrinsic value of subjective states and the value of individuals, but such a connection requires further value premises, namely, equality and identity. It could be argued that pleasure might rule out equality in certain cases, and thus as a qualification of it represents another value theory. Nor does the value of individuals follow from that of pleasure, without further qualification. The pleasure must be over a lifetime, or a single pleasure at the beginning of a life might justify killing, surely an odd ethic. Norton argues that intrinsic value is not dependent on other humans;[330] thus his refusal to extend intrinsic value to animals unless based on subjective valuing by humans seems inconsistent. As Norton himself argues that there can be "objectively better" values as the result of transformation, why can't the environment be defended along the lines of "objectively better" values? True, the value of an object is not the same as "objectively better" values, but the latter might be used to construct a case for the latter.

In terms of the criticisms Norton has made of individual positions on non-anthropocentric value, I fail to see the disanalogy of corporations to species. Norton argues that corporations represent the interests of individual members, but legally the corporation is an entity separate from its members. As in a species, individuals come and go, but the corporation/species endures. Species also are "born" or appear at a point in time and "die" or go extinct by analogy with individuals. That they have an interest in not going extinct is revealed in Darwin's "struggle for existence." Thus the interest of species can be described, given that interests are morally relevant, a point that I contested earlier, and which Norton argues against in his whole book. In short, Norton's arguments against demand value are an excellent argument against both utility and economic demand value. Thus criticisms of non-anthropocentric intrinsic value that presume such a theory of value are at best inadequate, at worst inconsistent.

The problem of the intrinsic value of the environment has been outlined through a consideration of some of its major spokesmen. The devaluation of the natural world in modern philosophy is coordinate with the increasing exploitation of wild nature for human use. Radical environmentalists have attempted to counter this trend with a notion of the intrinsic value of the nonhuman, and with the consequent moral standing and obligations of humans to the nonhuman based on such standing. Different approaches to the intrinsic value of the environment have been covered, from moderate, subjective theories (Norton) through more radical, relational theories (Callicott) to quite radical, objective intrinsic value theories (Regan and Rolston). The attempt to assign intrinsic value to nonhuman nature has been criticized on ontological, epistemological, and valuational grounds. In this section I have also examined the locus of value, the bearers of values, the content of intrinsic value (e.g., in Regan's "subject of a life,") and the "ontology" of values, from less to more radical theories. There is no consensus even among non-anthropocentric value theorists about the scope, locus, and character of nonhuman intrinsic value. Thus the issue arises of whether the approach from intrinsic value is fruitful, a point that will be addressed in chapters three and four.

The attempt to found a new, environmental ethic is radical only in certain respects. The movement has not generally pioneered new value theories or rethought the relation of values to moral obligation. What is new is the attempt to extend the scope of intrinsic value beyond human bearers, the question of locus of value in such bearers, and the questioning of the restriction of moral standing to humans, anthropocentrism.[331] Sometimes the scope of intrinsic value is questioned, as in Regan, sometimes also the locus, as in Callicott. The anthropocentricity of moral standing is questioned, however, by all environmental ethicists, whether their value theory is or is not anthropocentric. Each of the major non-anthropocentric figures I have discussed starts from established value theories and simply questions the restriction of value to humans. Even Callicott, who argues that extension of traditional morals will not do and calls for an overhaul of ethics, invokes Hume at crucial points. It would seem that any traditional value theory might, *ceteris paribus*, be modified for use as an environmental ethic, so long as it was compatible in most respects. This will be the premise of chapter five.

Another important issue is whether the notion of intrinsic value in consequentialist theories such as utilitarianism can generate obligation. In a sense environmentalism must be consequentialist in ethics as certain outcomes—preservation of species, of ecosystems, of individual animals—are the basis of the enterprise. The link between intrinsic value and obligation has already been questioned by two of our authors on different grounds, Callicott and Rolston. This is a problem for consequentialist theories on other grounds as well, a point I will examine in chapter four. I also hope to show that Dewey has put forward an excellent proposal on how such a link can be made.

PROLOGUE TO CHAPTER TWO

*The Setting of the Problem of
Pragmatism and the Environment:
The Critique of Pragmatism as an Environmental Ethics
in Taylor, Bowers, Katz, and Weston*

In this section, critics of pragmatism from within the literature of environmental ethics will be discussed and analyzed. I will discuss the figures who originated the controversy surrounding whether pragmatism can be used as an environmental ethic. Bob Taylor argues that John Dewey's naturalism is anthropocentric, not ecocentric, and thus cannot provide a sufficient model for environmental ethics; and that Dewey's is an instrumental view of nature, in which nature exists primarily for human exploitation.[1] Taylor argues that despite his naturalism, Dewey was really a social liberal in his approach to nature and that his naturalism was anthropocentric.

C. A. Bowers argues that what Dewey took to be objective, modern science, is actually culture bound. Since Dewey gave priority to the scientific viewpoint, he did not take alternative cultural interpretations of nature and the place of humans in it seriously. Further, Dewey the progressive had little use for tradition, including traditions of native peoples who have learned how to live in an environment. Thus the outlooks of such peoples, which form valuable paradigms for an environmentally conscious age, were ignored.

Eric Katz has argued[2] that pragmatism in general is inherently anthropocentric and subjectively oriented in value theory, that this results in relativism. Thus pragmatism cannot provide sufficient justification for the preservation of the environment or a basis for an environmental ethic. Katz argues that pragmatism is not only subjective but that Dewey's critique of intrinsic value undermines the entire project of extending the scope of intrinsic value to include animals, living things, species and the biosphere. Finally, Weston defends pragmatism as the basis for an environmental ethics despite its "subjective" theory of value.

Taylor as a Critic of Dewey

Taylor's article is in response to an earlier article by Chaloupka that emphasizes the significance of Dewey's philosophy for environmental thought.[3] The reading of Dewey as a technocrat misses the mark according to Chaloupka; Dewey's "social aesthetics" as a basis for an interrelationship with the environment is stressed. Taylor responded that Dewey is basically a social and political philosopher rather than a strict naturalist. Dewey's "human oriented naturalism" is concerned with construction of a uniquely human environment, not being "one with nature." "His naturalism certainly does not display an understanding of the human relationship to nature in any presocial, precultural or simply biologically oriented sense."[4] In other words, Dewey's naturalism is anthropocentric, thus nonhuman nature is viewed instrumentally. For Dewey, according to Taylor, "we must rethink our relationship with nature, but nature continues to be thought of in instrumental, anthropocentric terms."[5] He judges Dewey "insensitive" for stating that humans should "subdue" wild nature.

By way of qualification, Taylor admits that Dewey is not *anti*-environmental; nature is viewed as a teacher and even as an environment. However, Dewey's philosophy is primarily social rather than environmental, or, rather, the human environment is primarily a social one. Dewey is portrayed as a social liberal first and a naturalist second. Dewey is still within the Lockean tradition as both argue from an anthropocentric premise: "humans and their peculiarly human environment stand above the rest of creation."[6] Even if he believed that the "conventional dualisms of liberal theory" made it impossible to understand the correct relationship to the environment, he was simply unaware of environmental problems in the contemporary sense. Taylor remarks that "what is striking about both Dewey's liberalism and the more conventional Lockean liberalism is their shared lack of sensitivity to, and interest in, the types of environmental problems facing liberal . . . societies today."[7] Environmental thought is historically new and Dewey is a challenge rather than a basis for environmental thought, especially environmental political thought.

On the whole, Taylor's criticisms are one-sided, that is, they stress one aspect of Dewey's voluminous output while ignoring the overwhelming balance.[8] In the first place, Dewey's naturalism, by common consent, was influenced the most by Darwin. Darwin was also appealed to by Callicott, one of the more radically ecocentric figures. Dewey spends almost the entire first chapter of *Experience and Nature* explicating his sense of how humans and their experience are both "in and of" nature: derived from nature in the Darwinian sense of evolved from within it as well as living within it and experiencing within it. This is both anti-Cartesian and an implicit environmental stance as human experiencing is viewed within the larger context of a nonhuman natural environment. True, Dewey did not deal with contemporary environmental problems, any more than any of his contemporaries did. However, he came far closer to them in overcoming Cartesianism with his "web of relations," his emphasis on the social influences on individual subjects, and so on. The charge of anthropocentrism must then be attenuated: it is certainly not a subjective sort of anthropocentrism, and his orientation to nature, even human nature, shows where he believes humans should be centered. Indeed, Taylor himself quotes Dewey to the effect that nature is an "ally." He misses the significance of another of the quotes: "nature reshaping itself." The human impact on nature is within and part of nature. This can hardly be called anthropocentric. Human activity is also natural and its impact must be dealt with on that level, but this argues against a view of humans as detached from their environment.

Dewey's political thought is indeed concerned with human affairs, but this is the nature of political life, which concerns an exclusively human institution. No environmental philosophy argues that human society is the same as the relations in the wild. Approaching Dewey as if all he wrote were political philosophy or as if all of his philosophy was politically motivated is a distortion. His "instrumental" view of nature was not unique to his view of nature; Dewey called his philosophy both "instrumentalism" and "naturalism." Everything, not just nature, was viewed in instrumental terms precisely because of its relational place in a larger whole. This view is a more consistent development of a similar view in Callicott, Rolston, and other students of the ecology. Dewey came closest to articulating an environmental philosophy in the era before the rise of deep ecology, which focuses on the relations of the ecology as a model for environmental philosophy. I will use Dewey's thought as a paradigm for pragmatism in chapters two through five and consider the other pragmatists only briefly at the end. Dewey is the most naturalistic pragmatist, who considers our relationship to the environment most thoroughly. Further, he has the richest theory of value, and has integrated this into his naturalism.

Bowers and Environmental Education

C. A. Bowers has argued that Dewey, though seemingly constituting an excellent starting point for education in environmental issues, is inadequate as a "forerunner of ecological thinking."[9] Bowers, like Taylor, recognized that Dewey viewed the individual as inseparable from the environment. However he provides three main reasons why Dewey could not serve as a model for environmental education. The first is Dewey's "failure to recognize the culture-language-thought connection. . . ." According to Bowers, "that different language communities organize and experience 'reality' according to different root metaphors was not part of Dewey's awareness: thus he could not see his own epistemological preferences in terms of its culturally specific nature."[10] Dewey gave "privileged status" to "the scientific method of problem solving" and this constitutes an "unconscious" assumption. Further, "Dewey's anthropocentrism can also be seen in that other aspects of the environment are not given a voice . . . the deer, the trees and mountains are not the source of analogic knowledge."[11]

Second, Bowers charges that "the past, or what others term 'tradition' . . . was not viewed by Dewey as being a legitimate source of authority in guiding human practices and beliefs." Bowers argues that "for Dewey the essential attitude toward life is not that of preserving living traditions that have insured communal survival in a limited habitat, but an experimental attitude, framed in terms of the perspective of the single individual or social group and guaranteed by the belief that each scientifically grounded plan of action represents the outward expansion of human empowerment . . . and progress." In other words, Dewey the progressive had little use for tradition. Dewey is "forward-looking" in his view of nature and thus reads history from a progressive perspective. He represents a "modernizing form of secular consciousness."

Third, Bowers argues that the political process is one main means of instituting cultural changes. While it is acknowledged that Dewey recognizes the importance of the political process for modifying traditions, Bowers believes that for Dewey, "the only legitimate approach to the politicizing process involved the use of the scientific method of inquiry." Bowers acknowledges that "Dewey understood experimental problem solving as reworking and advancing patterns from the past" but was "unable to acknowledge that the scientific model of decision making is only one of the many valid forms of knowledge, and that valuing substantive traditions is not always a betrayal of human intelligence."[12] He argues that "there is no way to limit the politicizing process." Dewey was also "optimistic" about the "efficacy of the political process when guided by scientific reason and democratic practices."[13]

Most of the issues discussed by Bowers will be covered in succeeding chapters, in which I will show how Dewey's philosophy can indeed be used as an environmental philosophy. However, some of Bower's charges can be answered immediately. The charge that Dewey was either unaware of or insensitive to the role culture plays in human life and understanding is simply astounding. Bowers himself is either unaware of or has ignored Dewey's *Freedom and Culture*, a book that deals with those very issues for almost 200 pages. In the book, Dewey notes that "culture, which has become a central idea of anthropology, has such a wide sociological application that it puts a new face upon the old, old problem of the relation of the individual and the social."[14]

He continues, "with the intellectual resources now available, we can see that such opinions about the inherent make-up of human nature *neglected the fundamental question of . . . how their pattern is determined by interaction with cultural conditions.*"[15] Dewey was very much aware of cultural differences and of the influence of culture on the beliefs, practices, and the like of individuals.

Second, Dewey was not a critic of tradition as such. On the contrary, he argued that the source of morality lies in tradition, that is, in folkways or "customary morality."[16] The latter would never be entirely displaced, since there is no reason to totally revise it. Solutions from a culture that worked in the past should by all means be invoked for present problems. *It is only where customary morality breaks down or cannot be an aid in dealing with a new problematic situation* that "reflective" morality and what Bowers refers to as the "scientific method of problem solving" enter in. I will address this point in more detail in a later chapter. The point here is that a holist like Dewey cannot but include tradition. It may even have priority to a member of a culture. Yet tradition is not sacred: it is one important source of norms, but not the only one.

Dewey believed that, ideally, all the different cultural voices would be heard in a democracy. He was indeed optimistic on this score. However, democracy was more than a political term for Dewey. Democracy was both an ideal project and the method of social cooperation in all cultural affairs.[17] Dewey's rich political philosophy is beyond the scope of this book. But the idea that there was "no way to limit the politicizing process" implies a kind of total politicization: totalitarianism. Dewey was clearly opposed to totalitarian practices and theory and defended liberal democracy as opposed to totalitarian regimes of both the right and the left.[18]

Katz as a Critic of Pragmatism and Intrinsic Value

Eric Katz argues that the critique of intrinsic value in "pragmatism" undermines the whole rationale that environmental ethicists have been trying to develop within philosophy. Pragmatism is not a sufficient basis for an environmental ethic because "the values of pragmatism are inextricably bound up with human desires and interests."[19] "Pragmatism and environmental ethics must part company over the role of human interests in the determination of value."[20] Katz sees pragmatism as anthropocentric because it is based on human interests in value theory and only takes such interests into account in dealing with the environment. Anthropocentrism has resulted in a type of imperialism of humans over the balance of nature even though humans are within nature. Humans justify this human imperialism by the rationale that "human interests lie at the center of all value determinations," and "human good is the only determination of value."[21] Such anthropocentric imperialism based on human-interest values "destroys the value of the subjugated entity," the "particular complex of values" of the original entity, just as in cultural and military imperialism. Only human use is taken into account, that is, nature is treated instrumentally, not as intrinsically valuable. Pragmatism's form of utility is even considered egoistic or individualistic.[22] If this were used as a basis for dealing with the environment, the consequence would be relativism, in which ethics for the environment would be at individual discretion. Katz believes that a genuine environmental ethic would reorient value theory away from locus in individuals toward "species, systems and communities." Katz argues that pragmatism "places the value of the natural environment squarely on the *experiences* of human beings interacting with nature: the desires and feelings of human subjects."[23] Value is tied to human experiencing by pragmatism, which is inevitably subjectivistic, and this includes the value of nature. In sum, "pragmatic" value theory is subjectivistic and individualistic, based as it is on experience; egoistic, based as it is on individual desires and interests; and thus results in relativism.[24]

Pragmatism, therefore, cannot be an adequate basis for an environmental ethics. Katz questions whether an environmental *ethic* can be based on individual desires or interests or "favorable experiences." "Ethical obligations do not derive their force from favorable experiences," and may have force despite them.[25] Such an ethic is subjective and this is inevitably relativistic in Katz's view.[26] Katz regards this aspect of the value theory of pragmatism as unfortunate as pragmatism is otherwise well suited as an environmental ethic, for example, in its value pluralism and holism. Katz argues for overcoming dualism in ethics, contextualism, and value pluralism, pragmatic themes.[27] However, its anthropocentrism is unacceptable from an environmental point of view. "It is the basic policy of human civilization—even where that policy is unarticulated—to modify or to conquer the natural world, to subdue nature for the furtherance of human good."[28]

Katz even questions the model of intrinsic value as the basis for environmental ethics.[29] He argues that intrinsic value is tied too much to nominalistic ethics, the ethics of individuals, and that this is unsuitable for preservation of species and the value of habitats. Intrinsic value tends to focus attention on anthropocentric values such as sentience and rationality.[30] Such anthropocentric notions of intrinsic value are clearly not capable of justifying environmental policies. He argues for a smaller role for intrinsic value than its prominence in the literature of environmental ethics heretofore has suggested. This includes the idea that not all obligations are grounded in intrinsic value.[31]

Katz's critique of "pragmatism" is general, so it necessarily has regard to trends, not individual figures. However, this generality seriously distorts issues of value in pragmatism. For one thing, the issue of intrinsic value has been internally debated within pragmatism. The more idealist wing upholds a strong notion of intrinsic value, based on aesthetics.[32] It is Dewey who is the great critic of intrinsic value, and he calls himself an instrumentalist not a pragmatist. Thus a critique of "pragmatism" cannot be made on Dewey alone with respect to this issue. Second, there is no figure in classical pragmatism of whom I am aware who literally upholds the "interest" theory of value. On the contrary, Dewey was an acute critic of it, while the other pragmatists were not in sympathy with it. Katz probably means that human interests are being upheld by the anthropocentrism that he ascribes to pragmatism, but this is also misleading. True, Dewey and James base their philosophy on experience. But they are all attempting to get away from experience based on the Cartesian subject, to overcome Cartesian residues. Even the issue of grounding philosophy in experience is not unanimously assented to by pragmatists. Schiller, the English pragmatist, bases his system on values, not experience, while Peirce's version of pan-psychism is not based on an experiencing subject confronting an object, and C. I. Lewis begins with logical analysis.

Finally, Dewey does consider what is desired in his treatment of values but contrasts this with "the desirable."[33] His nuanced treatment is hardly reducible to desire simple, but considers the desired in contrast to the desirable, that is, demand or wants in relation to what should be wanted, desired, or demanded. There is a notion of obligation in Dewey's ethics that goes beyond relativism. This point applies *a fortiori* to the other figures. In sum, pragmatism is variable on the issues raised by ethics, rejects the interest theory of value, and is not grounded in the Cartesian subject. Katz's analysis of "pragmatism" as some sort of ethical monolith is a critique of a straw man.

Weston's Defense of Pragmatism

Katz and Weston engaged in a debate over the issue of intrinsic value in relation to the environment and pragmatism.[34] Weston defends pragmatism on just the lines for which Katz criticizes it. He agrees that pragmatism rejects the means-ends distinction and that this is for the good for environmental ethics.[35] He agrees that based as it is on experience, pragmatism is subjectivist, but argues that the more extreme consequences drawn by Katz do not follow, for example, relativism and anthropocentrism. "It does not follow that only human beings have value; it does not follow that human beings must be the sole or final objects of valuation."[36]

Weston considers and rejects the arguments for both intrinsic value and the intrinsic value of the nonhuman. He includes a discussion of the logic of means to ends as the rationale of intrinsic value as delineated by Moore.[37] According to Weston, Moore argued that intrinsic values must be self-sufficient, abstract, and have special justification. That is, they must be valuable apart from a relation to a valuer, be ultimate in value, and grounded in being, divine commands, or human subjects.[38] While Weston agrees that there can be an aesthetic experience of nature that is isolated in some sense, he argues like Dewey that a value experience is not a justification for intrinsic value.[39] Further, values are concrete and plural, not abstract or monistic.[40] Finally, he rejects any notion of an ultimate justification of values. Values can keep referring beyond themselves in an ongoing instrumental relation similar to that outlined by Dewey in his critique of intrinsic value. Values do not end in some abstract intrinsic value, although they may obtain "completion" when connected with a subjective human self. "Intrinsic values are urgently sought at precisely the moment that the instrumentalization of the world . . . has reached a fever pitch."[41] Ultimately, the whole issue of the intrinsic value of the nonhuman arises out of Cartesian metaphysics, which the pragmatists are almost unanimous in rejecting. Values are concrete, plural, and sometimes inconsistent.

Values are interrelated and interdependent for pragmatism. Weston portrays pragmatic value as the web of relations that the mutual instrumentality of value creates. This web undercuts intrinsic value; there is no self-sufficiency of value.[42] He argues that instead of grounding in intrinsic value, "we must learn to relate value" in assessing choices. "On this model there is no ultimate reference or stopping point simply because the series of justifications is ultimately, in a sense, circular: to justify or explain a value is to reveal its organic place among our others."[43] Rolston's attempt to tie intrinsic value to nature "weakens" nature as a source of value, by isolating it from the rest of value. Weston would prefer to "assimilate environmental values to our other values"—that is, to bring them into the existing web of relations.[44] This web is constituted by desire in Weston's view, "interlinked and mutually dependent," forming a "constellation" of desires, beliefs and choices. The web articulates the mutually supporting role values have to one another. Grounding is not needed, only "the relation of these values to other parts of our system of desires, to other things that are important, and to the solution of concrete problems."[45]

Values are contextually situated and arise in concrete situations and circumstances. Thus the justification for Alaskan National Parks should not be in terms of some abstract intrinsic value, but because of the specific values of the parks, their fragility and exceptionally unspoiled wildness. Weston gives an exposition of how values are historically and culturally constituted over a long period of time, and the effect of this on philosophic ethics.[46] Ethics is firmly embedded in a concrete social context and this fact is becoming a "consensus" among several major figures in ethics. Weston wants to strengthen environmental ethics by giving it time to evolve; it is now at the originary stage culturally and its effects on the culture of the West are uncertain. Weston wants enough time to assimilate environmental ethics so as to make nature an "organic part of our lives."

It is clear that Weston is using Dewey as a representative pragmatist at least in part, based on his critique of intrinsic value as such. Weston has not addressed the issue of what means are means toward, what instruments are instruments of. A chain of causes can regress to infinity, a chain of means cannot. He might respond that they are means toward other means, but this does not end the chain, qua chain. If they are means in a web, are they a means to the creation of the web? The end answers the ultimate question of why; a web of relations serves no purpose without such an end and does not justify its value apart from human exploitation. Further, as he suggests, there can be value experiences satisfying in themselves, such as contemplation of pleasing landscapes, which "disconnect the frame of reference in which value questions . . . arise."[47] This undercuts the interconnectedness of all value and suggests experiences valuable in themselves. It almost seems as if he appeals to intrinsic value in some passages, but a subjectivized view of intrinsic value, for example, in his statements that the experience of nature can awaken respect for it, or that parks are "exceptionally wild." This is close to the view that value is relational: our experience of the aesthetic is valuable in a sense disconnected from other values.

Weston has argued that intrinsic value is rooted in Cartesian metaphysics.[48] What he misses is how Cartesian his own analysis is, with value grounded in an experiencing subject and that subject's desires. Thus there is grounding of value, but in the subject rather than in nature. It is also in conflict with the relational view he also espouses. No relation to a subject is needed for such a web in the wild, as Leopold, Callicott, and Rolston have argued. Further, other environmental philosophers have criticized the relation of subjects and objects as exacerbating the dualism of human and natural and thus as the rationale for separation of value and exploitation of the wilderness, notably Dewey. Weston's argument that intrinsic value is of little practical import in concrete value decisions applies even more to an extraneous relation to a metaphysical subject. More, the ultimate reference to a self that "completes" values is egoistic. This is not the prevailing view in pragmatism and it is difficult to see how an environmental ethic can be constructed out of it. His claim that pragmatism is pluralistic in values contradicts his remarks on desire as the common element in valuation: value is any object of desire in this view, a view criticized by Dewey. Weston's arguments against the univocity of 'value' in intrinsic value theories apply as well to his theory of desire. Again, a web of relations could conceivably exist outside of anyone's desire; it is unclear why the referring back to desire is needed. His distinction of grounding and justification is thus also unclear since he argues against grounding of values in favor of a web of relations constituted by desire, that is, as grounded in desire. In sum, Weston has interpreted pragmatism through Cartesian blinders and thus has fallen into the trap he claims for his opponents.

2

DEWEY'S NATURALISM

> To see the organism *in* nature, the nervous system in the organism, the brain in the nervous system, the cortex in the brain is the answer to the problems which haunt philosophy.
>
> —Dewey

In the recent literature of environmental ethics, certain criticisms of pragmatism in general and Dewey in particular have been made. The argument has been that certain features of pragmatism make it unsuitable as an environmental ethic. Bob Taylor argued that Dewey's naturalism in particular is anthropocentric because it concentrates on human nature.[1] Eric Katz asserted that pragmatism is an inherently anthropocentric and subjective philosophy. In this chapter, both of these views will be challenged in the context of Dewey's naturalism. Dewey does not speak for all the pragmatists, and thus Katz cannot be answered entire, with respect to pragmatism as a whole. However, I will argue that Katz's charge misses the mark with respect to Deweyan pragmatism.

Dewey's himself referred to his overall philosophy as "naturalism," that is, as a "naturalistic metaphysics." Even his method includes a reference or relation to nature: the method of "empirical naturalism." This methodological nomenclature is important for it points to the connection made between empirical experience and nature. Dewey does not conceive of experience as outside of nature, but as within nature and thus of nature or natural. "A naturalistic metaphysics is bound to consider reflection as itself a natural event occurring within nature because of traits of the latter."[2] Reflection can be made on nature because nature is in some sense the origin of reflection and reflection is within the scope of the natural. Human nature is an outgrowth or development within nature. Dewey's project is to show the continuity between human and the rest of nature, not to emphasize the difference. He explicitly rejects the Cartesian model of an isolated self that is detached from the rest of nature or society. As such a subjective outlook is the basis for anthropocentrism, Dewey's view is also a rejection of strong anthropocentrism. Dewey has not separated the human from the nonhuman but, on the contrary, incorporated human value into organic processes of growth in general. To anticipate, he uses the general model of the organism in an environment for human nature. His use of growth and organism make his philosophy extremely important for environmental ethics. In effect, he has articulated an "environmental ethic" of a sort since the relation of an organism to the environment is the basic model defining the value situation. His view is naturalistic while maintaining values. The human is part of the natural, and this redefines the project of philosophy. "Our problem is to see what objectivity signifies on a naturalistic basis; how morals are objective and yet secular and social."

The Metaphysics of Nature

Dewey regards metaphysics as "cognizance of the generic traits existence." As what exists is natural for Dewey, metaphysics is the study of the general characteristics of nature: it is a specialized study of nature as a whole. Dewey is a consistent naturalist: even art is in nature and reflects natural activities. As Dewey rejects transcendence in any form, that is, existence beyond nature, as unknown by scientific methods, there is no distinction of metaphysics and nature except in terms of levels of treatment. However, Dewey rejects the identification of nature with the subject matter of science, a point of view that he describes as intellectualism.[3] Metaphysics is the study of nature in general, the traits of all existing natural processes. For Dewey a pluralistic universe of change is closest to the "empirical facts."[4] Humans do not stand outside of such changes in their lives; on the contrary, their own lives include such changes. "The reality is the growth process itself; childhood and adulthood are phases of a continuity. . . ."[5] Thus growth is real, an existing process, which is as much a part of nature as mechanical causes and effects. Growth as an end is also real, for growth comes to an end and this is a natural process, universal, as far as is known, for all things. Such growth is "the way of nature in man." Insofar as growth is a value, this value is both in nature and a natural good. "At every point and stage . . . a living organism and its life processes involve a world or nature temporally or spatially 'external' to itself but 'internal' to its functions."[6] Humans are in and of nature: they live in a natural environment and their own organic and bodily processes are natural.

A book could be written on Dewey's concept of nature, which is both rich and detailed. To cover his theory of nature in detail would be beyond the scope of our topic. However, a short precis of his main ideas can be presented. First, nature is what we are "in and of."[7] For organic creatures, space is not simply the "place in which we live," but also an environment. Being in nature also involves a system of relations, to other natural events or processes.[8]

Second, nature is the existence of events, processes, and histories. This dynamic view contrasts with existence conceived as being and coming to be, the Greek view.[9] Nature as change cannot primarily refer to fixed being; for the existence of change is then left out of the account. According to Dewey, every existence is an event; event is a "character" of nature,[10] and the complex of events constitutes nature. Change is a process marked by events that form a history: a sequence from a beginning or origin, a span of time in existence, and an end in which the process of change is finished because what existed has gone out of existence or changed into something else. The history from beginning to end is a process of change that marks all natural events, and includes the relation of cause and effect as such a process of change. Dewey believes that this latter relation is marked more by continuity than by discrete stages.[11] Effects are also causes, just as ends are also means, and there is no finality to the chain of causes and effects, only a stream of events.[12]

As events are in time and form part of a process of change, marked by beginnings and endings, they are historical. A historical process forms an irreversible sequence of events with an origin and an ending. Dewey argues that everything natural is historical[13] in the sense of going through such a process, whether quickly or slowly. "When nature is viewed as consisting of events rather than substances, it is characterized by histories, that is by continuity of change proceeding from beginnings to endings."[14] Events have finality, not in the sense of teleology,[15] but in the sense that they are part of a history: the process comes to an end or finish. Natural finality is the termination of an event.

Growth, one of the most important overall values for Dewey, is a process of change, marked by phases, but also by continuity. The phases of change come and go, while growth is more continuous. Growth as continuity is stable, for it continues through all the changes. The phases are precarious, for, like youth, they are lost in the process of change. Dewey argues that the generic traits of nature must include consideration of the "precarious" along with the "stable"[16] for these are traits that mark all growth, and even changes of inorganic nature such as geologic processes. Dewey argues that the "uncontrolled distribution of good and evil is evidence of the precarious, uncertain nature of existence."[17] Good and evil, cause and effect are viewed as "part of one and the same historic process, each having immediate or esthetic quality and each having efficacy, or serial connection."[18] As everything in time is natural, or part of a historic process, nature includes both immediate qualities and connections to other events. The precarious is connected to the stable as event, whether of one existence (an object and its changes of qualities or characters) or events in relation (e.g., cause and effect). In nature both elements, stability and precariousness, are present as a mix.[19] Any account of nature must include these as elements or it is not true to nature, but only a partial account. Dewey also regards the qualities of events as real and includes these in nature. Nature is regarded as the "affair of affairs" in which each affair is regarded as having its own qualities, which begin and end in time.[20] Nature is "immediate and final" in all its qualities: like good, they are just there, and that all that can be said of them. But they are no less parts of nature than are natural goods, which may be taken as a species of quality.

The precarious in nature is contingent[21] on causes and other events. However, because certain qualities are precarious, there is room for novelty in nature. Novelty is possible because of potentiality: it is the precarious possibility of natural processes. Nature includes potentialities as the condition of novelties and therefore the possibility of the novel: what never existed before. Possibility or potentiality[22] is linked to novelty: Dewey argues against the view in which nature can have "no hidden possibilities, no novelties or obscurities. . . ." His view includes "bringing about" new objects of value through activity: creation of good inventions. As natural processes can bring the novel about by themselves, for example in the form of the evolution of new species, novelty as such is natural.

Finally, nature has levels,[23] in which more basic processes combine in novel ways to produce higher, more complex processes. The evolution of life is such an emergent overall process, in which "higher," that is, more complicated organisms emerged from "lower" ones. Experience is an emergent character with greater scope in higher than in lower organisms. Thus it is a novel and perhaps accidental by-product of natural processes, the evolution of life. However, Dewey thinks that it is because experience has evolved as part of the world, that it can interact with the world.[24] Levels of nature do not take us out of nature, but, on the contrary, put us more acutely in touch with nature at the highest level of experience. The world of experience in which we live—that is, "the nature in which we still live"—includes the mix of the complete, orderly, and recurrent with the possible, ambiguous, and singular, which marks nature in general. Our lives take place in nature, and this includes experience of the natural world as a natural event within it.[25] Natural life includes certain kinds of experience that dualism would as soon leave out of nature, for example, feelings of satisfaction, and the prizing of them. Enjoyments in actual experience "represent the consequences of a series of changes in which the outcomes or ends have the value of consummation or fulfillment."[26] Enjoyment is a natural process in which a series of events has reached a culmination in experience. The enjoyment and the prizing of it is "in and of" nature.

What is the place of human nature within nature? Is there such a thing as a human nature at all? Dewey argues that human nature is not "forever the same." Basically he rejects the idea of rationality as the specific difference of humans, arguing that reason is something that is achieved, not native, and that it is not always operative in humans. What is common to human nature is a raw set of potentialities, referred to as "native human nature." These native potentialities include impulse and certain instincts,[27] which humans as "living creatures" possess. They also include organic processes that take place without conscious intervention or awareness, such as digestion. The native potentialities are incomplete as they stand, however, for they are always culturally mediated. Dewey notes the lesson of history is "the diversity of institutional forms and customs the same human nature may produce and employ."[28] Culture is diverse, so the form the native potentialities take depends on cultural upbringing. Cultural mediation results in the formation of habits in the form of customs that channel impulses and instincts into socially approved paths. "Man is a creature of habit, not of reason nor . . . instinct."[29]

Humans are able to adapt to culture because they are adaptable or plastic creatures. Their "nature" consists in potentialities or possibilities allowed by the structure of impulses and instincts, which must be adaptable to be of use.[30] Humans do not have a fixed nature, but an indeterminate one. The plasticity of human nature, what Dewey calls the "plasticity of impulse," allows adaptation to change. Individual humans are described in terms of the "plasticity and permeability of needs and likings " and Dewey argues that "this ambivalent character is rooted in nature. . . ."[31] In summary, Dewey notes the "uniform workings of a common human structure" amid the diverse social settings. But this structure does not exist apart from cultural mediation: it works through social custom and habit.

Since for Dewey human nature is part of nature, plasticity or indeterminateness is an attribute of organic nature. Human nature is *in* nature *of* which it is a part.[32] Human beings are primarily "living creatures," and life is a "trait of natural organisms,"[33] as well as the link between physical nature and experience. Dewey pointed out the isolation of humans from the rest of nature in modern, dualistic thought, "whose effect was to isolate the individual from his connections both with his fellows and with nature, and thus to create an artificial human nature, one not capable of being understood. . . ."[34] Humans are actually connected to the balance of nature as living and organic beings. Dewey rejects the classic species differentiation of humans as the "rational animal" in its modern Cartesian form, a detached thinking thing confronting a mindless, mechanical, extended thing. "Thought, intellect is not pure in man, but restricted by an animal organism that is but one part linked with other parts, of nature."[35]

To be sure, Dewey recognizes "levels" within nature. "The distinction between physical, psycho-physical and mental is thus one of levels of increasing complexity and intimacy of interaction among natural events."³⁶ However, the mental is still within nature as an event in and of nature: caused by and continuous with natural processes. The evolution of higher powers is a fact of organic nature; it was a character of evolution prior to the rise of humans. Though Dewey does distinguish human nature from the rest of nature, he locates this specific nature in an organism continuous with the rest of biological nature.

Humans must adjust to their environment or to prevailing conditions as much as any other living creature. The processes of life necessary to human survival are comparable to those of any organism, for example, nutrition and digestion. The mental functions, though perhaps higher in some respects, require such living processes in order to function. Life is seen as the link between mental processes and the rest of nature, as mind is in a living being that in turn is in a natural environment. Dewey wants to reconsider the "connection of conscious life with nature" because recognition of this relation undercuts the dualistic separation of mind and body, experience and nature. Mind is always found "in connection with some organized body," which in turn "exists in a natural medium" with which the body connects adaptively.³⁷ Just as plants require "air, water and sun," so do humans require sustenance from the natural environment, which involves interaction with it.

Dewey argues again and again against the separation of human nature from the rest of nature. "Man can be envisaged as within nature and not a supernatural extrapolation."³⁸ Again, "the human situation falls wholly within nature. It reflects the traits of nature; it gives indisputable evidence that within nature itself qualities and relations, individualities and uniformities, finalities and efficacies, contingencies and necessities are inextricably bound together."³⁹ So closely does human nature reflect nature in general that traits of human existence can be seen as traits of nature itself. Thus the distinction of human nature is no different in many respects from the distinction of any species nature: it is not privileged. While the changes brought about by human action are recognized, as is the emergence of a special function in the form of social communication, these are in turn shown as consistent with any favorable adjustment by a species of its environment. They are within the range of organic nature: Dewey views humans naturalistically. Taylor's and Bower's contention that Dewey is more interested in human nature than in the balance of nature⁴⁰ simply flies in the face of his treatment of humans as completely within nature and partaking of natural processes in their very lives. Taylor and Bower read Dewey as interested in the human subject because of an anthropocentric view of nature. However, Dewey's project is to bring humans back into nature by showing the continuity of human nature with nature and undercutting the dualistic metaphysics of the subject. To read Dewey as a subjectivist not only ignores Dewey's explicit repudiation of the metaphysics of the subject and his extensive arguments against it, but also his behaviorism. It ignores Dewey's frequently repeated analysis of human activity in terms of living "organisms" adjusting to their environment, that is, as being *primarily continuous* with other living things. To depreciate this is to ignore the entire thrust of Dewey's attempt to overcome the Cartesian legacy of the isolated subject[41] through his "naturalistic metaphysics."

"Experience and Nature," the title of Dewey's general work on metaphysics, involves the attempt to connect nature and experience. This title might seem to connect Dewey with the tradition of the experiencing subject, starting with Locke and including many of the British empiricists. This is misleading, however, for Dewey redefines experience in the process of working out its relation to nature within a naturalistic metaphysics. The metaphysics of experience for Dewey is within nature, not removed from it in a solipsistic, private world. For metaphysics is concerned with "the nature of the existential world in which we live."[42] It is not concerned with a transcendent realm or a special realm of a subject different in nature from the balance of nature and thus as detached from it essentially. The world is the world in which we live, in which our organic processes of life go on, including processes of mind, which require attachment to the body. "That to which both mind and matter belong is the complex of events that constitute nature."[43]

Dewey treats mind as part of nature, and not a distinct sort of substance, by redefining experience in terms of activity. Experience as activity is an event continuous with and in interaction with other natural events, differing only as a kind. Since mind experiences in some sense, it is brought along into the process of natural events along with experience. Dewey acknowledges that experience in the older sense has been used to mean private "mentalistic" experience. However, he argues that it can also "be used as a generalized equivalent of behavior."[44] Experience is not mere perception for Dewey or the totality of perceptions in a stream of consciousness. With James,[45] he argues that it includes life, activity, desire, imagination, and other elements and that the mental part cannot be separated from the total experience. It is experience of the world we live in and includes the activity or behavior as an event of experience. There is no division of the act from the material on or through which it acts, of subject and object. The experience includes the outer as well as activity connecting the organism with its experienced surroundings. From a metaphysical point of view, "things interacting in certain ways are experience . . ."—experience must involve interaction within a field of events. Interaction of events generally constitutes the essential elements of any experience. The link to human experience is that this is how they are experienced.[46]

Dewey argues against the separation of experience and nature. "Experience is of as well as in nature. It is not experience which is experienced but nature—stones, plants, animals, diseases, health, temperature, electricity and so on."[47] The separation of nature and experience by the earlier empirical tradition has a corollary: the idea that nature could be complete without experience or by excluding experience. However, even if experience were mentally defined (e.g., as perception), Dewey's arguments would apply. Perception is of other objects such as stones, plants, and so on. Dewey argues that experience does not shut out nature, but discloses reality in it. What is experienced is a "manifestation of nature" and thus an accurate character of natural events. Experience in the cumulative sense involves the "growing progressive self-disclosure of nature itself." Although reason maybe utilized as a method, in investigating nature we "start from and terminate in directly experienced subject-matter."[48] In turn, experience is dependent on nature. "Experience when it happens has the same dependence upon objective natural events, physical and social, as . . . the occurrence of a house."[49] Dewey means dependence both in the sense that it is experience of events and that the experience is itself an event within the natural world. Perception is the end product of a natural sequence of events, a process.

Experience conceived as outside of or opposed to nature would be "an experience which experiences only itself."[50] However, we observe things as objective or as external: as "there." Science would be impossible if it could not observe events as external. This constitutes evidence for the validity of experience, as to be skeptical of all of science in the light of its many successes and predictions would be practically impossible.[51] Further, as experience itself is a natural event, its own activity is a study in nature. Experience is not removed from the rest of nature, but an event within it. This event can be studied like any other, such that qualities of nature are revealed.

How can experience provide a valid connection to nature? It is crucial to understanding Dewey that experience involves both the body and the mind: body-mind. I noted previously that mind is always found in connection with a body. Mind was part of a historical sequence in which greater levels of awareness arose within a body or living organism. Dewey calls this view the "emergent" theory of mind in which three levels of interaction are distinguished. The first is the merely physical, the second the realm of life, and the third is mind.[52] The evolution of mind was part of a natural series of events in which more complex forms of interaction arose in response to a natural environment. Mind emerged after a sequence of historical events within a body through which it operates: body-mind. Thus there cannot be a division of mind from body, as in the Cartesian separation of mind from matter. "Body-mind designates an affair with its own properties."[53] Neither body nor mind can be separated from one another; the mind is not isolated from nature, for it is attached essentially to body, by which it experiences. In "body-mind," the body represents "the cumulative operation of factors continuous with the rest of nature."[54] Mind is also a product of social interaction: it is not formed in isolation but in response to a social and cultural milieu. "Mind can be understood in the concrete only as a system of beliefs, desires and purposes which are formed in the interaction of biological aptitudes with a social environment."[55] Mind is formed through interaction of the body with the environment in experience: of instinct and impulse mediated by social factors and the natural. Mind in general arose as a historical event within nature; individual minds are formed by interaction with a natural and social environment. The interactions of mind are "a genuine character of natural events."[56]

Consciousness, experience, and mind do not constitute a separate realm of being, then, a distinct substance, but are part of the totality of events that constitute nature. Indeed, it is the culmination of nature: "the manifest quality of existence when nature is most free and most active." Experience includes a mind that in turn is connected through a body to an environment, in that it lives in the larger world of nature. Nature is a series of overlapping events in which relative place is an essential identifying element: being *in* to be of. Mind must be in body to live, life in a suitable environment, and so on. Mind is not mind unless it fulfills such imperatives of place. Mind is not only within nature, but also of nature, as it arose by natural processes, acts through them, interacts with them, and is constituted by them.

The metaphysic of value is also natural for Dewey. "Goods . . . are as they should be; they are natural and proper."[57] Good is normal; it is evil that is problematic. As good is natural it is part of the world of nature, that is, it is in the natural world.[58] This includes the good of an organism, which is part of the natural world: the organism cannot be separated from its natural environment and remain alive. Its good is a good within the natural world and requires interaction with the natural world used as an instrument. In accordance with his Instrumental theory of value, Dewey argues that "nothing in nature is exclusively final," and everything can ultimately be used as a means as well as an end. The Instrumental theory is a naturalistic theory of value, in which good is part of natural processes, that is, has a beginning and end in time or a history. This includes the good of valued objects, which eventually decay or lose value; as well as the good of activity, which comes to an end as the final act in a series or process. Neither of these senses of good, that attached to things or to activities, is different in essence from other natural processes. Natural ends, in the sense of terminations are ends of events.

Dewey can thus argue for the naturalness of moral and esthetic values. If experience "presents esthetic and moral traits," then these also belong to nature. The traits of experience are not simply in nature, but revelatory of nature. Esthetic traits in particular "characterize natural situations as they empirically occur."[59] Morals are also judged "objective" on a naturalistic basis. Morals as primarily connected with activity are a mode of experience; indeed, experience as activity must incorporate moral norms. "Morality depends upon events, not upon commands and ideals alien to nature."[60]

Esthetic and moral traits are "found," not simply created or inferred. By way of qualification, however, morals and esthetics are viewed as "secondarily natural."[61] Morality incorporates and "adjusts" the natural environment in "bringing about" ideals or creative realization of novelties. This process allows 'ought' to become 'is': an ideal ought is actually brought about.[62] It is natural because it incorporates natural ingredients, although the end product is not naturally occurring. An artwork uses brush, canvas, and paint derived from naturally occurring things. But the artwork is only secondarily natural, as it is not a direct result of natural processes, but reflects creativity and deliberation. These in turn are in and of nature, but their end product is an alteration of the natural in favor of what is prized. They are concerned with what is not in existence except by effort and change.[63] The organism alters the environment to suit it: its activity realizes or preserves what it prizes. This activity alters the naturally occurring environment, but is still within the potentiality of nature. Otherwise, it would not be possible. Potentiality as a trait of experience is also a trait of nature.

Dewey noted the tendency of estheticians to confine good in itself to the human experience of beauty, "isolated from nature," in "minds independent of nature."[64] Dewey argues that on the contrary, value is in nature. "The foundation for value and the striving to realize it is found in nature. . . ."[65] Intrinsic value marks a certain stage of the process, its consummation. "Values are naturalistically interpreted as intrinsic qualities of events in their consummatory reference."[66] A consummation of experience is its peak as a process, that which has a "quality of immediate and absorbing finality." For Dewey, it is natural to have consummations in experience. Since they are what is prized or valued, humans are "naturally more interested in consummations than in preparations" although he cautions that these are precarious. Dewey has thereby tied value to all the general traits of nature: event, end, process, quality, and precariousness.

Moreover, values are part of an ongoing process of experience as activity, and thus within, part of, and dependent on a natural environment. The process of deliberating on means to ends involves an evaluation of cause and effects in the world, a natural process. It may seem like a contradiction to argue that values are consummations in experience, with a sense of finality, and also that they are means, and that nothing in nature is ultimately final. However, experience is itself a process marked by events with different significance. Values are significant events, which stand out in the course of some processes. Values are indicated if a part of the process attracts attention and is absorbing. They may also be the culmination in the sense of marking the end of the process as success, for example, the completion of a project or the actual achievement of a novelty. This is not the end of life or activity, as Dewey stresses. However, it may mark the end of a certain direction of activity and the start of new activity. The value culmination may mark finality of a type, as the consummation of experience. However, this is not ultimate finality, as the process of activity as a whole goes on: consummations are part of a larger process.[67]

Because value is within nature, fixed ends outside of or external to a process are subverted. According to Dewey, this is the true ethical import of the doctrine of evolution.[68] Fixed ends subordinate present change to a future goal, a model common in the philosophical tradition. However, ends are part of an ongoing process of change, not outside of that process. Evolution has meant continuity of change is a corollary part of this process, part of natural history. A merely stable world would have no consummations of experience, no peaks, or valued, if precarious moments. "To be good is to be better than; and there can be no better except where there is shock and discord combined with enough assured order to make attainment of harmony possible."[69] The process of adjustment to the natural environment is not always smooth and, when harmony is restored, the experience may seem good to the organism. The organism can then resume normal activity in the ongoing process of growth. Dewey notes that growth, the "sole end," is not confined to children, but is an ongoing process even for adults. It includes developing experience to widen it and thus continue the process of growth.

In summary, the metaphysics of value in Dewey is naturalistic. Dewey notes that some value theorists erect a value realm separate from or isolated from "natural existence." The problem with this approach is the relation of the two posited realms. For Dewey this approach is a mistake and the problems it raises are arbitrary, pseudo-problems. Value in general is part of natural processes or inventive continuations of them. Even intrinsically valuable culminations of experience are within and part of natural processes. Since for Dewey a quality within experience is a trait of nature, intrinsic value is a trait of nature—that is, *nature is intrinsically valuable*, at least in part.[70] Consummations of a process mark the value of natural events, which may include natural objects as goods whose realization in existence marks the culmination of an activity or process. Dewey, far from denying the intrinsic value of nature, argues that the experience of intrinsic value or consummation of experience is a trait of nature. It is not, however, extrinsic to and foundational for activity but part of an ongoing process within other natural events. Dewey has provided an alternative to foundational metaphysics that is more naturalistic. Thus he might think that the problem of the intrinsic value of the environment as it is usually approached is misconceived, for it presumes that a foundational model is the only model. Yet the holism of environmental relations and the delicate interrelations of parts does not lend itself to a foundational model. Environments are more like an ongoing process than a foundation. I will return to this point in chapter four.

II

Just as human nature, experience and value are within nature for Dewey, so are they "in and of" a specific environment. They require an environment to function and reflect this environment in their very constitution. Activity in the form of conduct, habits, and other behavior is oriented toward, takes place in, and interacts with a specific environment. "All conduct is interaction between elements of human nature and the environment, natural and social."[71] The environment is not simply a locus and medium of activity; due to "impersonal forces" it can provide a "stimulus" to activity. Knowledge, belief, and the like are an effect of the workings of the environment in interaction with natural impulses.[72] For example, the impulse to fear trains is mediated by an altered environment, which includes trains.[73] However, these alterations do not make it any less of a natural environment as a whole.

In turn, the environment itself changes historically through the very process of activity within it. It is in flux: altered by the interaction of the activity within it precisely to reflect value. The organism alters the environment to enhance valued outcomes and bring about what is prized. Thus the environment reflects amelioration over time. Further, the environment is highly specific, despite the generalization reflected in the term. "There is no such thing as an environment in general; there are specific changing objects and events."[74] Humans and other organisms find themselves in specific situations of climate, rainfall, and so on and must cope with these to survive. "Man is a being who responds in action to the stimuli of the environment."[75] The problematic situation that gave rise to valuation is within and partly constituted by a specific environment.[76] Having to cope with cold, for example, arises in specific environments as a problematic situation. Dewey argues that normally the environment is in balance with human and other organic activity: organisms are reasonably well adjusted to the environment. However, problematic situations do arise and formulation of goals represents the attempt to adjust to the new circumstances of the environment, the "means of effective adjustment of a whole set of underlying habits."[77]

Adaptation of organisms to a specific environment is not confined to humans. "Life denotes a function, a comprehensive activity, in which organism and environment are included."[78] All of life must cope with the challenge of an environment. Interaction with the physical environment connects humans as organisms to other living things. All organisms create an altered environment, not just humans. "What the organism actually does is to act so as to change its relationship to the environment. . . ."[79] Plants root themselves by penetrating into the soil, taking nutrients from it and leaving behind their by-products. This may not seem like an alteration compared to human technology, but is considerable by comparison with barren landscapes. More complex organisms, of course, make greater changes in the environment. These are melioristic, creating more favorable and valued conditions for the organism, conditions it prizes and cares for. What is required is "effecting that permanent reshaping of environment which is the substantial foundation of future security and progress."[80] In turn, living things have to adapt to changes in the environment that reflect organic activity upon it: organisms must be flexible to survive.[81]

Dewey's stronger claim in this context is that mind is shaped by the environment, that it is formed in relation to nature and society. "Different customs, established interacting arrangements, form and nurture different minds."[82] Mind is not autonomous in this view, solipsistically isolated and alienated from a mechanical and meaningless natural environment, but is shaped by the environment around it, particularly cultural customs. The environment, both social and physical, is the source of thoughts and feelings.[83] Mind is not only within nature; it is shaped by nature in the form of environmental influences. The subject is formed, in part, by the molding of personal habits in child rearing. The main means for humans is language, which is a social medium. "Psychic events . . . have language for one of their conditions."[84] Failure to recognize that thoughts, for example, are expressed in language, and that language was acquired in social interaction, has lead to the solipsistic, Cartesian view of an isolated subject. Dewey argues for a different model of mind based on its social origin through language. "'Mind' is an added property assumed by a feeling creature, when it reaches that organized interaction with other creatures which is language, communication."[85] The social environment permeates the inner as language: the inner reflects the outer in language and meaning.

Humans are not regarded as somehow out of their natural environment by Dewey. On the contrary, they are connected with it in activity. Activity involves using physical events as causes to effect needed or valued changes: as means. This process involves both the organism and the environment in interaction,[86] the organism acting in and on the environment in which it dwells. Experience is in and of an organism in an environment that shapes experience. It occurs under special natural conditions, "in a highly organized creature which in turn requires a specialized environment."[87] The mind can interact with the world "because mind has developed in that world," not distinct from it as an autonomous subject. Mind "uses the structure which are biological adaptations of organism and environment as its . . . only organs."[88] It works as a constituent element of a mind-body or organism in an environment congenial to it and functions in adjustment to its environment. "In every waking moment the complete balance of the organism and its environment is constantly interfered with and constantly restored."[89]

Humans and other organisms are intimately connected to their natural environment for Dewey: there is a continuity of the human, the organic, and the natural world. This continuity is the basis for pragmatism as an environmental ethics, by extension.[90] The issue is whether existing Western philosophies are capable of extension and suitability for an environmental ethic or must be completely discarded as too subjective or anthropocentric. I submit that Dewey's naturalism is not only suitable but also paradigmatic, for the environment is not excluded from human nature, as it is for subjectively based theories, but is crucial to human life. Desire, for example, does not involve a psychic event removed from natural surroundings. It involves "tension" and "effort," which is marked by an "active" interaction of organism and environment.[91] Any living thing must maintain a connection to the natural environment or die in Dewey's view; life takes place entirely within an environment. Humans as much as any other living thing require a connection to the environment to function as natural beings. "We live mentally as physically only *in* and *because* of our environment."[92] Human mental activity is one with the environment and the behavior of an organism is within a natural environment. This environment is where activities take place and also shapes the activities. Valuation arises in the interaction of an organism with its environment, a natural process in which valuation is constituted. In turn, the environment is actively shaped by the interaction to create more suitable conditions for the organism.

Taylor's argument that Dewey concentrates on human nature because of the active shaping, for example, in technology, ignores the melioristic shaping by other organisms. Although Dewey does analyze human nature, the latter is not separated from the rest of nature, but within it. Taylor may also have been put off or misled because of Dewey's conception of a social environment. He may have overlooked the fact that the social environment is itself within the physical environment for Dewey,[93] as well as common to humans and other gregarious species. It would be unrealistic to ignore the fact of human alteration of the environment, but this is in many respects continuous with other alterations as events of organic nature in general. Plants create soil as much as grow in it. The humus that is a constituent of soils is the by-product of the breakdown of plants. The channels that rain flows through when it goes into the soil are created by the growth of plant roots. The plants generally form a "root ball" around their roots, incorporating a portion of the soil as their environment, which in turn holds the soil in place during rains. The plant is "in and of" the soil and actively transforms it to make it more suitable for growth.

Dewey's naturalistic model could be the basis for a different approach to environmental ethics, one that avoids the problems of subjectively based theories. Environmental ethics are required because the environment is a condition of human life, our "home." The environment is not separated from the human, but part of the natural human context because there is continuity between human nature, the organic, and the natural environment. However high humans have arisen, they are still organisms who need an environment in which to live, act, and flourish. Activity is a natural event of an organism whose consummatory phase is marked by value. An action is simply one more event in the natural world, but it is melioristic to the organism. Valuing constitutes a natural process inseparable from the environment, a process of ongoing interaction. Value is what is naturally good for an organism, and is used as an instrument in its life processes. I will return to this point.

III

Behavior and activity are not confined to humans, but characterize organisms in general. It might seem a stretch to attribute behavior in the sense of conduct to animals. However, animals do exhibit behavior, instinct, and habit and this is the common ground between them as natural organisms and humans. Dewey's project is not to completely separate human nature, as I previously argued, but to include it in nature as a whole. It is not a misreading of Dewey to include humans as organisms since he refers to "human organisms."[94] To be sure, value involves more than just "vital impulses" because it also includes such elements as deliberation, evaluation, and so on. But the bearer is the individual[95] organism, not subjective states of humans. Organisms are active or the locus of habit, behavior, and activity. These traits are common to all organisms, thus not anthropocentric. If this reading is correct, Dewey cannot be interpreted as anthropocentric, either in his metaphysics of value or in what bears value.

This reading is also supported by Dewey's treatment of psychology in terms of behavior and "body-mind." The elements of mind are expressed as a kind of behavior; the body-mind acts or behaves in a natural environment. The mental is not distinct from the physical or the body but combined with it in habit, impulse, and conduct. As Dewey puts it, habits without thoughts are "futile," while thoughts without habits "lack means of execution."[96] Habits involve mind in the use of intelligence as a tool for more intelligent adaptation, but mind is not distinct from habits and behavior. It functions through them and thus is an element of the common function. "The dynamic force of habit taken in connection with the continuity of habits with one another explains the unity of character and conduct, or speaking more concretely, of motive and act, will and deed."[97] Dewey argues that habits are required for both mental and moral acts and this reveals much about his view of mind. Mind is made evident from "mental acts," that is, a kind of *action*, which includes behavior, custom, conduct, and so on. Mind is a function of an acting organism manifest in acts that are intelligent. The mental element is present but unified within the action. This is "mind-body" or the intelligent action of an organism. The activity of the organism is "psycho-physical" not a reflection of and consequent on the mental alone as a distinct substance, thing or factor.[98] The psychophysical also includes the conjunction of "need-demand-satisfaction," that is, of specific kinds of mental acts such as desire, feeling, and so on. These are also expressed in action or activity of the organism.[99] The "soul" is described as "qualities of psychophysical activities as far as these are organized into unity."[100]

Katz and Weston[101] have described pragmatism as a "subjective" standpoint or as supportive of subjectivity, whether in psychology or value theory. However, Dewey explicitly states that his use of the term experience is "not subjective" and that his approach is "wholly contrary to such a philosophy."[102] The concepts of "mind-body" and of the "psycho-physical" are meant to overcome the dualism of mind and body, of subject and object, which is the legacy of Cartesian philosophy in the modern period. He argues against a dualistic psychology in which mind is opposed or removed from an actual natural "environment."[103] Rather, mental events are included in acts or activity within the environment. The good attached to the 'mental,' including aspects of mental such as feeling and desire, are expressed in and as an element of actions. Dewey is not arguing that activity is mindless, but rather that including intelligence in the activity gives it superior value, while a mind detached from activity is without issue, and thus of little value.[104]

The "subjective" elements of mind-body are included as part of a natural process. Mind and matter are characters of "events," thus part of the natural course of events. They are often essential elements of this process, but not their foundation, either in acting or in value.[105] Mental functions represent emergent elements of organic history that have a value to the organism over and above the merely physical. But they are included within the organism as valuable elements in its interaction with the environment, as instruments for coping. "Mind denotes the whole system of meanings as they are embodied in the workings of organic life...."[106] It is not distinct from the organism and its activity. "Habits formed in the process of exercising biological aptitudes are the sole agents of observation, recollection, foresight and judgment: a mind or consciousness or soul in general which performs these operations is a myth."[107]

Katz and Taylor have not made their case in charging Dewey with being "anthropocentric." I have tried to show this in the metaphysics of value, in which value is treated as within nature, not attached to the human subject. The bearer of value is the organism in an environment, not some inner human faculty or psychological state, a point I will cover in more detail. The environment is considered essential to the life and activities of this organism and Dewey explicitly recognizes that it has a special claim. A further consideration is that Dewey is a value pluralist who recognizes not a single end, but a variety of "natural goods." He would like us to "advance to a belief in a plurality of changing, moving, individualized goods and ends...."[108] These are not limited to subjective human states or even to the human organism, for they include objects, functions, such states of the organism as health and even ideas. Value thus extends to both the environment and the organism as the primary bearer of value. Indeed, in the notion of "bringing about" new objects, a fluidity in the scope of value is recognized, for new objects of value increase the scope or extent of the valuable.[109] Nor is value conceived as a relation of objects to (human) subjects. Not only is subjectivism the "opposite" of Dewey's approach, but it is left almost entirely out of his account. Relational theories of value, that is, as between a subject and object, are criticized repeatedly, and the relation is even denied in the form it takes in such theories. "The 'relation' in question is understood to be plural . . . involving a variety of space-time connections of different things, not singular, and . . . the connections in question are across spaces, times, things and persons. . . ."[110] Dewey's theory extends pluralism from kinds of values to the many sorts of connections involved in a holism such as that recognized as the web of relations in a habitat or environment. Persons are only one element in this web. If this is anthropocentric, I do not know what a non-anthropocentric position would be.

The Bearer of Value

Philosophies of value are often metaphysically grounded in an "ontology" of value. Dewey, however, rejects any notion of fixed or enduring being. Existence is connected to events and processes, not substances, as I tried to show in the last section. Another term is therefore requisite to stand for that which has value, that to which value is attributed as an adjectival quality. Dewey's arguments for value as primarily "adjectival" and not a noun implies that it characterizes something. I will use 'bearer' because this is a fairly neutral term, free of unwanted metaphysical associations.[111] The bearer of value has value in some respect but this implies nothing about the bearer as such except that it is capable of having or bearing value. Value is primarily attached to activity and secondarily to objects for Dewey, as activity is the basis or cause for "bringing about" new goods and keeping old ones in existence through acts of "caring for" and the like. Dewey has a primarily active theory of value as manifested in acts of prizing, caring for, and artistic creation. Desire and interest are defined in terms of behavior or conation and thus come under this active characterization.[112] As Dewey conceived experience in terms of behavior, experience was interconnected to valuing, since both are behavioral in some respect, namely, valuing is a quality of intelligent experiential behavior. Insofar as such behavior is an event in nature, value is the inherent quality of such events.

Dewey's use of "experience" might seem to tie him to the metaphysics of experience, or dualism of subject and object. This is not the case, however. Dewey was one of the first to attempt incorporation of the behavioral view in psychology into philosophy and morals. Experience is not that of a subject, but a type of behavior. Dewey is an empiricist but of a new type, in which experience is viewed as behavioral. However, he has not avoided the Scylla of introspection only to fall into the Charybdis of physicalism. Behavior refers to "animal life processes," that is, it is defined in terms of biology not mechanism. The "field of value" is concerned with such life processes as "selection-rejection," and "maintenance," that is survival and the choices requisite to sustaining a healthy life.[113] Ultimately, morality is concerned with life as a whole. In this view of value, Dewey connects with the vitalist and organic tradition in ethics. However, he interprets life with his own twist, in which the growth that marks all of life becomes a moral standard of a sort.

What is the bearer of value in Dewey? Because value is primarily attached to activity in Dewey, it might seem as if activity is the bearer. Activity is connected by Dewey to habit, conduct, and behavior.[114] Since organic activity consists in behavior, modified by past behavior or habit and also by impulse, activity is the activity of the organism. It is the organism that selects, behaves, and has habits. Dewey would be the last to imply that behavior or habits could exist as activities apart from the organism who behaves, who acts. On the contrary, habits are defined as "organic modifications."[115] Value phenomena originate in "biological modes of behavior." Value is borne by an organism whose activity has a "positive direction of change." The organism is the bearer of value. The activities and habits of the organism have instrumental value for the organism as allowing it to live and cope with its environment. The organism itself has value as the locus or center of growth and life, the only true ends for Dewey. The organism changes in the process of growth: organisms are a special kind of organized process. These changes can have a positive or negative direction, which is the primary sense of value in Dewey.

Habits are also the criterion of the objectivity of value essential for a scientific ethics: the "psychology of habit" is described as "an objective and social psychology."[116] The success of the organism is objectively evident *as* result and this involves the entire process from initial problematic situation to successful (or unsuccessful) outcome. The good is "objective" in terms of consequences, i.e. success or failure in achieving goals and linking causes to effects. Dewey notes that "value expressions" are employed as "intermediaries" to bring about the desired change from present to future conditions. Successful coping is good in the sense of achieving the consequences, of not falling short. The organism must function well in the environment: a functional view of good, which is objective. "Habits are like functions in many respects and especially in requiring the cooperation of organism and environment."[117] Good as objective is an attribute of activity, which results in successful functioning in the environment as an ongoing process.

The primary locus of value for Dewey is in the individual. This is consistent with the bearer of value being the organism, which is an individual; and connected with behavior, conduct, and activity, which are singular events involving individual agents. Dewey states that "every act, every deed is individual."[118] Dewey also remarks on the unique good of the situation. However, there is also a sense of activities being connected in a longer process with prior activities being instrumental to later ones. Further, completed activities may be useful still later as instruments. Thus there is also a notion of the value of the process as a whole as well as the connections and transactions involved in intelligent behavior. Many of the elements or factors involved in the situation have their own good, for example, the tools may have instrumental value along with the activity that utilizes them, the environment in which they are used, the end at which they aim and the social consequences of the activity. Good is plural, and thus there may be more than one kind of locus.

Individual bearers and activities are primary loci of value, but the web of relations required for activity also has its place and value. Dewey argues for a holistic view and against nominalism or any individualistic approach in ethics (e.g., that of the conativists) that it is anti-social.[119] He emphasizes the many "connections and transactions" required for valuation as opposed to the single relation of subject and object in relational theories. Further, general notions of value have instrumental value in a situation. These are arrived at by induction from individual cases of value, but still have a distinctive value as tools for solving problems.[120] There is also a notion of species value in the notion of natural kinds, for example, in the classification of "such natural goods as health, wealth, honor . . . friendship . . ." and moral goods.[121] Dewey is aware that a value bearer must be individual in the primary sense. However, the individual is in a web of relations that connect it in vital ways to larger wholes and to specific kinds. The value relations may be more general than those attached to just the bearer. I am not arguing that Dewey was not a nominalist in one sense: he clearly is. However, with his rich notion of values, there is the recognition that the value realm must extend beyond the individual for the individual bearer to survive. The bearer of value is not always the sole locus of value.

The Scope of Values

These issues have a bearing on the scope of value, the more quantitative extent of bearers. Is value restricted to humans or is value of larger scope in Dewey? Does it include higher animals, lower animals, plants, all organisms, and life? Does the environment have value? The biosphere? An attractive landscape? All of nature? Clearly, all these can have instrumental value insofar as they lend themselves to activity in furtherance of growth. The question then becomes whose growth? That of humans or also that of other species? It might at first seem as though value were to be confined to humans, as Dewey speaks of the control of "human" conduct by values, especially conduct that is not basically instinctual or habitual.[122] The latter might seem to cover animal behavior and thus remove the latter from the sphere of valuation. However, Dewey also remarks that value is not confined to a "peculiar class of things. Anything under the sun may come into possession of what is named by 'value' as its adjective."[123] As I noted, organisms are value bearers, and animals and plants are organisms. Thus a case could be made for the value of organisms of all kinds based on Dewey's theories. This case can be built from activities as well, since, as Dewey argues, we cannot know subjective states, the basis of subjective value theories, except as they are exhibited in behavior. However, all organisms exhibit behavior, not just humans. The science of animal behavior is based on observing and interpreting animal activities by analogy with human. The issue is whether other organisms are alike enough in the relevant respects to warrant greater moral consideration than a mere tool or instrument. Dewey would of course say that a tool should be cared for, as it may be needed in the future. However, there is also the notion of growth as the end and value attaching to life overall, and both of these are more than mere instruments.[124]

Dewey does remark that caring for by animals is "quasi-valuational," for example, that exhibited by the parents of young animals.[125] Value accrues to "life processes of selection-rejection" and this would include organisms on all levels, which exhibit decided preferences of food, habitat, mates, and so on. "Animals, including man, certainly perform many acts whose consequence is to protect and preserve life. If their acts did not on the whole have this tendency, neither the individual nor the species would long endure."[126] Dewey's tendency to include the human in the larger sphere of nature extends to his treatment of life. Nonhuman organisms also act to preserve life and are little different from humans in this respect. Their acts have preservation of life, including the life of the species as an overall consequence. This is a biological imperative, as it were, that humans share with other organisms. Nor is preservation of life the only behavior with larger consequences that animals and humans share. Dewey notes that "an animal given to forming habits" is one with needs and that this results in the formation of "new relationships with the world about it."[127] Habits, needs and changing relations to the environment are organic, not confined to humans. More, animals are believed by some to be able to think because of the similarity of their physiological organs to those of humans.[128] If physiology is any evidence, animals and humans must have similar functions, experiences, and responses.

Dewey is at pains to show the connection of mind to body as mind-body and that this is within an organism that is in turn in an environment. Humans are not peculiar in this respect, but typical of organisms in general. "The physiological organism with its structures, whether in man or in the lower animals, is concerned with making adaptations and uses of material in the interest of maintenance of the life process. . . ."[129] Human adaptation to its peculiar environment is of a piece with other life in maintaining its life. Habits are ongoing instruments of this process, ways of adapting whose value endures. Organisms "tend to continue a characteristically organized activity . . ." by utilizing conserved past energies to adapt subsequent changes to their needs. Activity is organized for survival by organisms and humans are an example of, not an exception to, this process. "Thus with organization, bias becomes interest, and satisfaction a good or value and not a mere satiation of wants or repletion of deficiencies."[130] The origin of value lies in its contribution to the survival of a species' way of life. Dewey ties this process to organization of organisms, not to humans alone. Interest is not conceived subjectively, but organically, as part of the life process. If organisms are value bearers, it is difficult to see how nonhuman nature can be excluded from the scope of values and thus some moral consideration.

This non-anthropocentric view is borne out by other texts in which need, satisfaction, and sentience are treated within the larger scope of organisms, not the narrow scope of human value. Feelings separate the organic from merely physical nature, not the human from the rest of nature. Dewey argues for "an organism in which events have those qualities, usually called feelings, not realized in events that form inanimate things...."[131] Sentience is a mark of life, not just of human life. The divide is extended to need and satisfaction as well: needs and their satisfaction separate organisms, not humans, from physical nature. "Empirically speaking, the most obvious difference between living and non-living things is that the activities of the former are characterized by needs, by efforts which are active demands to satisfy needs, and by satisfactions."[132] Humans act as if livestock, pets, and crops have needs that require satisfaction, in feeding and caring for them. Wild organisms survive by tending to their own needs. "Plants and nonhuman animals act as if they were concerned that their activity, their characteristic receptivity and response, should maintain itself."[133] Survival is the condition of life for all organisms, not only humans. Life is the goal of all instrumentalities of value: value for life. The value of health is not confined to humans, but is both universal for organisms, and subject to objective verification.

Dewey does not deny a specific difference of humans and even an emergent qualitative distinction. Human good differs from the pleasures of an animal.[134] Further, "the ability to respond to meanings" differs in humans and animals, although animals can respond at some level. Nor do animals think, at least insofar as thinking requires specifically human languages.[135] However, language developed in a social environment and both the former and the latter are exhibited by some nonhuman species—in herd behavior, animal cries, cross-pollination, and so on. Language is an emergent character of natural events with continuities to the rest of nature. "Continuity is established between natural events (animal sounds, cries, etc.) and the origin and development of meaning."[136] The emergence of language was within natural potentialities exhibited in more modest form in animal communication (e.g., growls and distress cries). Humans are still animals despite qualitative differences that are exhibited by all organisms. "Qualitative differences, like those of plant and animal, lower and higher animal forms, are here [in language] even more conspicuous; but in spite of their variety they have qualities in common which define the psycho-physical."[137] The denial of the animal within the human is false to the facts as "man began as part of physical and animal nature." Human evolution as a distinct species is not a transcendence of nature but its continuity.[138] Insofar as value is continuous with nature and with the behavior of other organisms, it is plausible to argue that nonhuman organisms are within the scope of value in the relevant respects for Dewey. This might seem like an optimistic or exaggerated reading with selective quotations. However, even if this were the case, Dewey has provided a very good basis for extending the scope of overall value, the value of life and growth as ends, to animals, plants, and the biosphere. Very little extension from Dewey's existing theories is needed to provide such a basis; less extension is needed in his case than in competing theories.

Summary and Conclusion

Dewey was above all a student of Darwin; he was aware that evolution is open-ended and thus history and nature cannot be read as a prelude to the ascent of humans. Evolution does not have a fixed end; Dewey remarks on the "conceit" of judging the whole universe by human desires and dispositions.[139] Far from being anthropocentric, Dewey treats humans in common with other organisms rather than in terms of their specific differences. Humans are in nature, creatures of habit not of detached "mentalistic" faculties. To read Dewey as anthropocentric is to ignore his entire philosophy of nature, the framework of his whole philosophy. Human nature is indeed specified, but Dewey argues that it cannot be understood outside of its natural environment. His project is to bring human nature back into nature as a whole. This includes the naturalization of value as quality: value is also part of natural processes. Dewey argues against the dualistic separation of humans from nature that has led to the view of humans as somehow opposed to nature, of nature as a hostile force or enemy to be conquered. Humans for Dewey are not separate from nature but are an organic element within it.

Organisms are the primary bearers of value since they act, behave, and are capable of organized activity, the primary sense of value. Humans are organisms: they are animals who function more or less well in their specific environment. Humans share growth, activity aimed at "maintenance" or survival, and social behavior with other living organisms. These elements are relevant to value concerns as the ongoing focus of all instrumental values. Nutrition, for example, is a factor in survival and this is common to all organisms. The locus of value is primarily individual, that is, tied to individual organisms. However, Dewey's value pluralism also has a place for the value of species or natural kinds, of general values and of the whole.[140] Since the bearer of value is the organism and values are conceived pluralistically, a wide reading of the scope of values as including all of life seems justified, in more than a merely instrumental sense.

The root of the problem of the intrinsic value of nature is metaphysical. Value was subjectivized following the Cartesian subjectivization of metaphysics, since value was systematically derived from metaphysics for the main tradition of Western philosophy. This subjectivization involves the devaluation of nature. Dewey's pragmatism is a countermovement in both metaphysics and value theory in which both the subject and value are brought back into nature, as traits of nature. In this process, nature is revalued, since value is part of natural processes. Intrinsic value is included in the natural process as a part of it. However, intrinsic value is not the foundation of all value, especially in the form of singular, fixed ends that transcend the process. Rather, they are dependent on such processes. Dewey's philosophy is a corrective to the trend in modern value theory and ethics of subjectivizing value and basing obligation on subjective (intrinsic) value. Value is brought back into nature as a natural quality of events and other natural processes. It is as "objective" as any other natural element within experience.

Contrary to Katz's and Weston's reading of pragmatism,[141] I would argue that Dewey is anti-subjective, whether in metaphysics, epistemology, psychology, or value theory. His project consists in the attempt to overcome Cartesian dualism and the dichotomy between subject and object, practice and theory, and mind and matter. Contrary to Taylor's, Bowers', and Katz's reading, Dewey is naturalistic, not anthropocentric. Nature is not simply human nature. The emphasis is on the continuity of humans and nature, with humans described as organisms interacting with other organisms in an environment that they are in and of. Humans and nonhumans share growth and development; Dewey treats these as a mark of organisms. Dewey's project is an attempt at a naturalistic, nonsubjective and non-anthropocentric ethic and philosophy in which humans are brought back into the environment. Dewey's philosophy is a rejection of anthropocentrism both in metaphysics and value theory.

3

DEWEY'S INSTRUMENTALISM

Dewey's philosophy has often been interpreted as undermining or eliminating intrinsic value. This issue has been raised in the context of environmental ethics by a critic of pragmatism.[1] In this chapter, I will argue that this is a misinterpretation. However, this misreading may be based on Dewey's Instrumental approach to value. I will examine Dewey's Instrumentalism, and his critical scrutiny of the means-end relation in detail. The issues raised by his analysis will be treated in relation to the question of how an Instrumental theory of values can contain any notion of intrinsic value. I will also briefly address the issue of whether his critique of intrinsic value undermines any attempt to establish moral standing based on intrinsic value.

Other philosophers have criticized the notion of intrinsic value besides Dewey. Some have rejected the notion outright, for example, Nietzsche.[2] Others have done so implicitly—in the transcendence of value in certain forms of idealism in which value is not intrinsic to an object or to being, but transcends objects. Still another is the noncognitivist theory of Ayer and Stevenson in which it is argued that value judgments are not a form of knowledge but merely express emotion or feeling. This is an implicit denial of intrinsic value as it denies that value judgments are facts and thus part of the nature, essence, or attributes of anything. Dewey's position involves the rejection of all of these approaches, a point I will develop and examine in some detail. Dewey is particularly critical of the Aristotelian model on this issue. However, despite his sustained critique of the notion of intrinsic value, Dewey ends by refining rather than abandoning the notion. Dewey does not deny the relation; on the contrary he incorporates it after critically qualifying it, that is, he transforms it.

Means and Ends

Different accounts of intrinsic value in the tradition constituted a problematic situation of itself to Dewey: for which account is to be taken as correct? If the accounts are contrary—value as subjective-objective, transcendent-immanent, anthropocentric-non-anthropocentric, and so on—is intrinsic value a coherent notion? If so, which position is best? Dewey saw the need for a major overhaul of the notion, which in many respects was little changed from Aristotle's time.[3] Indeed, he maintained that "the relation of valuation to things as means-end" is one of the major problems for value theory. He noted that the extrinsic-intrinsic value relation was often treated as coextensive with the means and end relation, but that the equation of the two relations constitutes a retention of absolutism, as the notion of something valuable "in itself," is a "denial of connections."[4] Dewey begins with the identity relation between means and ends, and intrinsic and extrinsic value, including the reflexive statement of intrinsic value, 'good in itself.' Dewey successively analyzes and refines Aristotle's theory of intrinsic value, and separates legitimate use of 'ends,' such as ends-in-view, from intrinsic value. Dewey's theory of value is not so much an abandonment of the notions of intrinsic value and ends as their refinement for a more scientific age.

In the first place, it should be stated what Dewey's theory is not. It is not a denial of ends or that ends motivate action. This is evident from the notions of "end-in-view" and the "ends of action." The former is defined in terms of "what it would be better to have happen in the future," as well as "projection of an end-object."[5] "End-in-view" is already a refinement of "ends," however, as it implies or connotes that the end must be within view; ends not in view are distinguished, a point to which I will return. In this context, the point is that not only are ends-in-view part of the resolution of any problematic value situation, they are invaluable if not indispensable for it. The end extends and enlarges our view of the act by putting it in perspective.[6] Further, ends define action in some sense: Dewey has a teleological, not a strictly mechanical interpretation of behavior in which action is undertaken for some end or goal.[7]

Dewey's position is not a denial of the connection of means to ends, although he considerably refines this relation, which is central to his theory. An "idealized object... becomes an aim or end only when it is worked out in terms of concrete conditions available for its realization, that is in terms of 'means.'"[8] The end is a distinct element defined in relation to means; this presumes and even reinforces the validity of the means to end relation. What is critically evaluated is the neglect of attention to "the means which are involved in reaching an end." It is the separation of means and ends, not their relation, that is critiqued. The distinction is not absolute but "temporal and relational." "The 'end' is merely a series of acts viewed at a remote stage; and a means is merely the series viewed at an earlier one. The distinction of means and ends arises in surveying the course of a proposed line of action, a connected series in time."[9] The end is the "last act thought of," while the means are intermediate acts, although they are immediate when viewed from the perspective of the situation. Thus the division is not of kind but of degree, or of judgment. In another respect, the end is a "name for a series of acts taken collectively . . . 'means' is a name for the same series taken distributively. . . ."[10] Each act used as a means is ordered in a determinate sequence leading toward the end, which organizes them into the correct sequence to achieve the end.

Nor, for that matter, is his theory a denial of the relation of means to ends in the sense of consequences. Dewey endorses this connection: "ends, in the sense of actual consequences, provide the warrant for means employed—a correct position. . . ."[11] The means to consequences relation is identical with the cause to effect relation; in this Dewey agrees with Aristotle and much of the tradition. Ends as consequences are the "termini" of causal processes. However, he separates ends-in-view from ends as results or consequences, a distinction that will be discussed next. Dewey also is not denying that deliberation on both ends and means is called for.

Dewey is not denying the sense of 'ends' as consequences or outcomes, nor the relation of ends-in-view to consequences. Indeed, consequences are central to his theory of value, to speak nothing of other areas of his philosophy. However, he is aware of the ambiguity of the term "ends" as meaning both goals prior to their being brought about; and also "ends" as consequences, that is, objectives that have actually been brought about. Both of these notions of end are also distinguished from "end in itself," that is, from intrinsic value, a point that will be covered later. He carefully separates the notion of "end-in-view" from end as consequence while retaining their relation.[12] "All processes have an end, not in the metaphysical or quasi-metaphysical (often called 'mental') sense of that word, but in the sense in which end is equivalent to result, outcome, consequence—in short, is a strictly descriptive term."[13] Ends-in-view are ideational but *consequences* are observable as an outcome, often an object that has been "brought about" or created. Indeed, Dewey argues that the end as a consequence or outcome depends on the conditions or constituents that bring it about.[14] Further, once a result has been attained, activity comes to an end in the literal sense and this cessation of activity or behavior is evident. The relation of ends-in-view to consequences is that ends-in-view anticipate consequences: they are "foreseen consequences," which are "projected as ends to be reached." Humans like some outcomes better than others and attempt to maximize those that are valued. "Attaining or averting similar consequences are aims or ends." Ideally, the end-in-view will match the actual outcome. However, Dewey argues forcefully for their differentiation. "They [ends-in-view] are in no sense ends of action. In being modes of deliberation they are redirecting pivots in action."[15]

From end in the sense of a consequence, which brings a particular activity to an end, it is evident that Dewey is not denying that an activity can be brought to an end or finalized. Indeed, he does not, with qualifications, even deny "final ends." Activity is not an endless, pointless, Sisyphean task. For there is "continual search and experimentation to discover the meaning of changing activity, [which] keeps activity alive, growing in significance."[16] The goals of action give meaning to present activity by defining and deepening it. Such activities do come to an end, as an outcome that is achieved or actualized. The outcome may be evaluated as good if it accords with aims and resolves the problematic situation. Further, ends-in-view can be "final" goods in the sense of the "conclusion of a process of analytic appraisals of conditions operating in a concrete case. . . ."[17] In other words, it is the last value formed, the final product of an evaluation of a situation. Thus it is relative to a situation, not beyond it as some ultimate end. The final choice in a deliberation is the final value in a logical sequence, not a final value in the traditional sense. However, there can be a finality to intrinsic value in the qualified sense that Dewey formulates. Health, for example is "no means to something else. *It is a final and intrinsic value.*"[18] Thus a sense of finality can belong to a value on its own. Dewey's main concern is to deny that this means it has no connections to the rest of activity, life, and nature.

Dewey is not denying overall ends, general ends, or even a hierarchy of ends. Overall ends are regulative in some respects over situational ends in the sense that the best course of action is that which harmonizes with values on a larger scale: growth, life, and a positive direction of change. Without overall ends, situational ends might be isolated from the rest of activity, and Dewey is very concerned to connect them. "Growth, renewal of mind and body . . . [and] harmony of social interests is found in widespread sharing of activities significant in themselves."[19] The activity has its own value but also is connected to larger scale values. Again, general ends are formed inductively from particular ends and in turn form a framework for evaluating particular ends. General ends provide "limits" for particular ends.[20] In some cases, a particular end may also be an instance of a general end. Finally, evaluation of ends in terms of better and worse has as corollary a hierarchy of ends, at least situationally. This is also indicated by the distinction between desire and desirable, enjoyment and enjoyable, and like distinctions. A fixed, transcendent hierarchy of values in the Platonic sense is denied but on the ground of its being fixed and transcendent. Of several possible ends, however, a hierarchy of their comparative value is evaluated during deliberation.

Finally, Dewey's theory is not a denial of intrinsic value,[21] only a major qualification of it. It is important to emphasize this point as Dewey has often been taken to entirely abandon the idea of intrinsic value, or criticize it to the point at which it is untenable.[22] I do not mean that Dewey accepts the traditional notion without question. On the contrary, he finds it "ambiguous" and far from being "self-evident." Dewey explicitly argues that the good of certain values is "of final and intrinsic value," for example, health in the quote above. Intrinsic value is located in other kinds as well, for example, in economics. Economic goods are frequently regarded as instrumental by the tradition; Dewey argues that they should be considered as intrinsically valuable and final as any others. In this sense, intrinsic value is used as a critical notion by Dewey to measure devaluation. Means are devalued as unworthy when they are separated from ends. "Anything becomes unworthy whenever it is thought of as intrinsically lacking worth."[23] Dewey actually expands the notion of intrinsic value to include the worth of means as an intrinsic quality of them. Action as a means may also have intrinsic value; "so far as a productive action is intrinsically creative, it has its own intrinsic value."[24] However, he criticizes the notion of intrinsic value as detached from means, as entirely without relations or meaning. Intrinsic value is not equivalent to a fixed end or an "end in itself" either. Further, the idea that as intrinsic a value is beyond criticism or evaluation in certain circumstances is unacceptable. This point particularly applies to knowledge, considered as a practical activity. Although the value of "pure" research cannot be doubted, the results of such research may also be applicable to solving practical problems.

An intrinsic value is a "consummation" of experience[25] but is not isolated from the rest of experience. Intrinsic value marks a stage of certain processes, their consummation. "Values are naturalistically interpreted as intrinsic qualities of events in their consummatory reference."[26] A consummation of experience is its qualitative peak as a process, that which has a "quality of immediate and absorbing finality." It is a consummation that is a part of a larger whole, whether a life, a series of activities, or other ends. Consummatory experience has its own value as a quality of it, but value in this sense does not make reference to some abstract, fixed standard. It is a consummation within experience, not outside it. As such, certain enjoyments and satisfactions are certainly worthwhile, but are also connected to other parts of experience. Present consummations may prove instrumentally valuable in the future as aids in determining what will prove satisfactory in similar circumstances. Thus they are not absolutely final, an "end in themselves," as they are part of ongoing activity. After the enjoyment is over, new activities will commence to which the memory of the satisfaction will be contributory. Intrinsic value is genuine, but is part of a larger whole, to which it has connections.

The end may be particularly valued by experiencing creatures; this is the culmination of experience. However, such valued endings are particularly precarious or unstable: they do not last.[27] In a sense their precariousness is what makes them valuable, since if they were constant, they would not be a consumation of experience, but part of its background: they would not stand out as good. Activity aims at preserving a good, caring for or prizing what is unstable; or creating more goods, bringing about. Caring for requires that attention be given to making the means of what is prized secure: its *sine qua non*. They are rendered more secure through control of changes in the process, that is, of means.

As Dewey defines experience in terms of activity and behavior, it is bound up with valuing as a quality primarily of activity. Experience as activity intersects value. Experience as natural includes the value of experiencing as a natural activity, just as the experience of value is a natural event. Value is of wide scope since both natural objects and activity can be of instrumental value. Anything natural can potentially be of instrumental value.

II

Since I have analyzed what Dewey's theory of ends is *not*, we can now examine what Dewey does state about "ends in themselves." First, Dewey's theory of ends is critical of the idea of timeless or transcendent ends, that is, as Platonic.[28] Since intrinsic means a quality "actually belongs" to something, it cannot be independent of space-time connections.[29] Because a quality cannot transcend that of which it is a quality, intrinsic value is historical, part of a historical process.[30] If the notion of intrinsic value is at all valid, it must be temporal, as a quality refers back to its "bearer" as a quality of it. Thus Dewey's own notion of overall ends must be interpreted as in time and space, not as timeless or transcendent. This does not necessarily imply ontology, as the bearer might be a subject, a psychological state of a subject, an experience, a whole, an activity, and so on. However, the relation first articulated by Aristotle of a quality referring back to a bearer of value may be essential to any notion of intrinsic value. This is because the quality must be a quality of something *as* a quality. Therefore it cannot transcend its bearer, except as an abstraction, not as an actual quality. Dewey is not denying intrinsic qualities, only non-observable extrinsic relations that establish some sort of basis for intrinsic or inherent qualities, including any absolute theory of intrinsic value as totally self-contained and without relations.[31] Dewey is criticizing theories in which value is unnatural or supernatural and infinite.

It is clear from his critique of transcendent ends that Dewey's theory is a critique of ends removed from activity, actual situations, and processes. Value is tied to the situation and to activity, even where it forms a consummation of them. The end of one activity is not the end of activity, only the "present direction" of activity. It is the beginning of a new activity by the same actor; humans, or for that matter living things, cannot cease activity altogether. Further, ends "arise and function within action. They are not . . . things lying beyond activity at which the latter is directed. They are not strictly speaking ends or termini of action at all. They are terminals of deliberation, and so turning points in activity."[32] Having distinguished ends-in-view from consequences, Dewey can now argue that ends-in-view are entirely relational,[33] for they only arise relative to a problematic situation and are designed to resolve it for the better. Normal activity impeded by circumstances is "the source of the projection of the end." Ends are selected to "liberate and guide present action out of its perplexities and confusions."[34] Dewey argues that this is the sole meaning of ends and purposes.

As ends are relational, Dewey can consistently criticize the notion of isolated ends, that is, ends removed from consideration of relations to overall values, ongoing activities, and consequences. "When ends are regarded as literally ends to action rather than as directive stimuli to present choice, they are frozen and isolated."[35] Isolated ends are the model for theories that posit ends-in-themselves as isolated from connections and hence unchanging or timeless. Dewey scores the separation of such ends (e.g., religious or esthetic ends) from the "interests of daily life." The separation involves a failure to view the end as a means to further consequences in an ongoing process rather than some sort of absolute finality in the course of events.[36] Ends may later be means, a point that was touched on in examining Dewey's notion of intrinsic value. They are part of an ongoing process in which they are now ends and later means to further consequences. Treating an end as isolated is to treat it only for the short term, not in terms of longer views and more ultimate consequences.

From his critique of "frozen" ends, Dewey can proceed to criticism of single, fixed ends, unrelated to circumstances. Dewey notes that ethics is "hypnotized" by the attempt to find some ultimate good that is a "single, fixed and final good."[37] This has provided a persistent model for ethics, in which fixed "ends in themselves" provide goals and regulate action. The result has been the separation of consideration of means from ends and of ideal from material goods as well as devaluation of the latter. Dewey argues, however, that "means and ends are two ways of regarding the same actuality," and that separating them has unfortunate consequences.[38] More, he denies that there are any "fixed, self-enclosed finalities." "Ends are literally endless, forever coming into existence as new activities occasion new consequences."[39] Ends-in-view are tied to the situation, and because the novelty of this circumstance is what called for the end as a goal, it cannot be based on a fixed end. New activities and changes call for new solutions. A fixed end is a single end, but this belies the plurality of ends indicated by different kinds of ends and even more of circumstances. Dewey states categorically that "there is no such thing as the single, all-important end."[40] Dewey judges picking out one end as "arbitrary," despite the fact that it is common in traditional morals. Justifying the means in terms of some singular, ultimate end reflects thinking in which it is "an end-in-itself," thus valuable "irrespective of its other relations."[41] The lack of relations, connections, and thus of overall value is problematic rather than the notion of intrinsic value simple.

Since a single, fixed end cannot coherently be a warrant for the multiplicity of activities in changing circumstances, it follows that a single, subjective fixed end cannot provide such a warrant. Hedonism for example, which is based precisely on such an end, is ruled out. Dewey also doubts that desires, interests, and "likings" can provide such a warrant, for this would put them beyond appraisal or evaluation. Desires, interests, and likings are problematic; the issue is whether they have merit, whether they are desirable. Circumstances, means, alternative desires and ends, and consequences must be taken into account, among others. In other words, conation must be subject to the entire procedure of deliberation, valuation, and valuing to be warranted.[42] Similarly, the intrinsic value of enjoyments ignores the question of the "regulation of these enjoyments." Dewey argues that "enjoyment becomes a value when we discover the relations upon which its presence depends."[43] The question is whether something is worthy of being enjoyed (cf. sadism); its relations must be evaluated before it can be adjudged intrinsically valuable. Nor are enjoyments the only kind of intrinsic values. Actions and means can also have intrinsic value, since this is a quality belonging to that which has it. Since what has value is often actualized as a consequence, and this is both predictable and objective, value is not confined to subjects or subjective states. Indeed, even enjoyments are regarded as more active by Dewey than in subjectivist theories.

Dewey's theory is a denial or critical of the notion that intrinsic value is identical to an end in itself. This point follows from his upholding a reformulated sense of intrinsic value while criticizing any single fixed end in itself. In separating intrinsic value from ends in themselves, Dewey made a major break with tradition, which tended to equate the two notions.[44] For Dewey, 'ends in themselves' is a "contradictory notion," since an end implies a connection beyond (i.e., prior to) itself, that is, the end of something. Thus it cannot be an "in itself." Intrinsic value as a property has a relation to that of which it is a property, the bearer, at the least. Thus it cannot be an end in itself if the latter notion is taken as nonrelational. As an example, Dewey notes that some enjoyments come at too high a cost. Thus their value is measured in terms of their cost, not per se, and thus not simply as an end in itself. As I previously noted, an intrinsic value in the sense of a consummation of experience may later be used as a means. Thus it is not simply an "end" in itself. Further, means may have the quality of value as "inherent," for example, a reliable tool.

Finally, and perhaps primarily, Dewey's theory is a critique of the separation of means from ends. Theories of intrinsic value as fixed, singular, transcendent, timeless, isolated, or separated from activity all involve a separation of ends from consideration of means, which Dewey regards as one of the major problems in value theory. It echoes the separation of theory from practice, intellect from instinct, foresight from habit, and mind from environment, which has marked the dualistic tradition in Western philosophy. Dewey argues extensively and vehemently against the separation of ends and means, since actual ends cannot be brought about except through determinate means.[45] This argument is aimed against theories in which ends are removed from the specific situation and the means requisite to that situation. Ends that do not involve such a relation are not relevant to it; they are misconceived. Despite the claim of a rational warrant by the tradition, the separation of ends and means in the form of ends that are detached from the situation by virtue of transcending it is not logically related to the specific situation. Dewey argues that their separation is "abnormal" and thus unwarranted; such ends are based on impulse or whim not deliberation over "existing conditions,"[46] and involve a misunderstanding of their correct relation to means. "Means are means; they are intermediates, middle terms. To grasp this fact is to have done with the ordinary dualism of means and ends."[47] They are the middle of which the end is the finality or end-in-view. Means are conceived in relation to a specific situation, as means intermediate between the problem and the solution. To separate them from ends involves making ends of means, as then the means would be the immediate end of the situation, the end in view.[48]

Dewey also makes specific criticisms of any approach that separates means from ends, for example, that such an approach involves a model that, as whimsical, is immature. It is to act on impulse, the mark of a childish character, rather than intelligently mediated desire and interests. Ends are not rationally related to the situation, nor thought through in terms of larger, long-term consequences but based on arbitrary desires of the moment. Dewey contrasts such models with the union of prizing and appraising, of desires and interests that are "matured and tested."

A further consequence is the devaluation of means, which, as separated from ends, are not valued as such. This point was raised earlier in the context of intrinsic value; to deny intrinsic worth to something is to devalue it. Dewey argues forcefully for the revaluation of means: means can have inherent value also in the sense of qualities.[49] The notion that means are "extrinsic values" involves a "contradiction in terms," if qualities are inherent. For example, the value of a tool is not extrinsic to it, or it would not inhere in the tool. However, Dewey also rejects the idea that 'inherent' is out of relation to anything else or completely timeless, separated from circumstances, and so on. Both ends and means can have inherently valuable properties but be formulated in a practical situation and thus stand in relation to it. Further, means can be prized and cherished on their own account; in this respect they do not differ that much from ends. "The idea that things as ends can be valued, cherished, held dear apart from equally serious valuing of the things that are the means of attaining them is . . . a fallacy in theory."[50]

The worst consequence of the separation of ends from means, however, is that the end is used to justify nefarious means. Although as we have seen, means have a "value of their own . . . 'intrinsic value . . .'" the logic of justification by appeal to intrinsic value alone rules out evaluation of means. The result is that the end justifies the means, which overlooks evil means.[51] Dewey pragmatically concedes that "certainly, nothing can justify or condemn means except ends, results." Nevertheless, he argues that consequences also have to be included: a single "liked" end does not justify other bad consequences, including consequences of using certain means. "Not the end—in the singular—justifies the means, for there is no such thing as *the* single all-important end."[52] He denies the that there can be any valuable ends detached from evaluation and appraisal of means.[53] Thus ends as ends-in-view receive further warrant from ends in the sense of consequences or results. By separating them, Dewey can be consistently both a consequentialist in certain respects and also argue against any ultimate end justifying the means.

Dewey should not be read as a critic of the Aristotelian notion that an end can later be used as a means; on the contrary, this model is central to his refinement of the means-end continuum. He remarks that the "sole alternative" to an arbitrarily selected end is that "desires, ends-in-view and consequences achieved" be appraised or *re*valued in terms of means to subsequent ends or consequences.[54] What is now an end or consequence of a series of activities used as means may later become a means itself, for example, invention of a new tool to do a specific job. Any end, including some satisfaction of desire, may be of value in a later situation as "a means . . . in further activity."[55] Dewey's analysis attenuates the paradox in Aristotle of ends later being means and intrinsic values becoming instrumental values by deemphasizing the finality of ends and the ultimate value of intrinsic value. Ends as consequences are only temporary ends of a direction or kind of activity, which later may be reincorporated into activity as means. Intrinsic values are not ultimate, but temporary consummations of experience, which are later instrumental to further consummations. By denying that there can be any absolute highest good, a summum bonum, or any transcendent, timeless, isolated, fixed end removed from activity, circumstance, or consideration of means, Dewey has removed the relation from the sphere of paradox. Dewey's examination of the relation is a critique of ends formed without regard for means or isolated from means.

For Dewey, the entire means-end relation is continuous with other activities, and thus part of a continuum in which means and ends succeed each other in time and in relation to circumstances and the resources available. In this "continuum of means-ends," means have their own, equal value, and ends can be of instrumental value without paradox. Because the end-in-view marks off a particular activity and acts as a "coordinating factor" of "subactivities," for Dewey there is no paradox in a "temporal continuum of activities in which each stage is equally end and means." Each activity finished is the means to the next; intermediate ends must also be taken seriously. Means also must be brought about, for example, the invention of new tools. Prior to bringing them into existence they are themselves an end-in-view or ideational end.[56] As I emphasized earlier, ends ought to be determined on the ground of the means entailed and often must be. Thus there is no great separation of ends and means for Dewey. "Means and ends are two names for the same reality."⁵⁷ Dewey recalls the story of burning a whole house to roast a pig as a satire on not considering means in relation to ends. He argues that the means in this story can only be adjudged absurd when the relation of ends to means is considered.

This continuity of means-ends extends to the value of the end, in the sense of consequences or actual results, as well: "the value of . . . something which . . . stands in relation to the means of which it is the consequence . . . hence . . . as mediated."⁵⁸ The intrinsic value of the end is not denied, so long as its relation to the means and activity with which it is continuous is recognized and that it can be subsequently reappraised as a means to other ends.⁵⁹

III

An instrument implies instrumentality for something. What are instrumental goods instrumental to or for? This question separates into two parts. One of these was previously covered, that is, Dewey's extended analysis of the relation of means to ends. The other, the end for which instrumentalities function, can be treated separately. It is clear that Dewey has a notion of an overall value: an "inclusive" or "comprehensive" good.⁶⁰ It is inclusive, as all immediate values must harmonize with it to some degree; it includes these particular values in an inclusive relation. Such inclusive values include happiness, growth,⁶¹ and life itself. An inclusive end *harmonizes* immediate, situational, and particular ends and gives them order.⁶² For Dewey, the sole end is growth; "growth itself is the only moral end." Thus "the process of growth, of improvement and progress, rather than the static outcome and result becomes the significant thing."⁶³ There are several elements introduced by this model. The first is an organic process: growth is organic and takes place over a period of time. Thus Dewey has connected instrumental value to the living organism, to vital processes.

Second, growth as a process does not have a "fixed" end: it is ongoing.⁶⁴ As an example, one can take health, the health of the organism. Dewey speaks of "recovering from sickness" as an "intrinsic and living process." He argues that activity for an organism such as humans does not culminate once and for all, but must continue in accordance with processes of life. "A continuous process is the end and good. The end is no longer a terminus or limit to be reached. It is the active process of transforming the existing situation."⁶⁵ Third, since the process is ongoing, it involves continuing activity. Growth is not separated from activity, but connected with it. Finally, if this process is in a good direction, then it results in improvement, progress, and amelioration. Life gets better as the environment is continually enhanced to favor those elements that have proven to bring satisfactions and other benefits. Thus the direction of activity is part of an ongoing process of development whose end is improvement, a better life.

Each good in a life is equally good: Dewey is a pluralist in values. The instrumentality of value extends to many different natural goods. Dewey argues against any one single end that "there are a number of such natural goods as health, wealth, honor . . . friendship, esthetic appreciation, learning and such moral goods as justice, temperance, benevolence, etc."[66] The naturally good seems to be implicitly distinguished from the moral in this passage, echoing a Hellenic distinction. Further, the natural goods are listed as kinds or species of goods, although this represents a logical classification rather than an ontological one, as Dewey does not argue for natural kinds.[67] They are a general classification of goods.

In summary, Dewey's treatment of the means-end continuum argues that ends can always be instrumental. Because ends can later be means, the distinction of ends and means is not absolute but relative to circumstances. "Actual consequences, that is, effects which have happened in the past, become possible future consequences of acts still to be performed."[68] What is an end in one situation may be a means in another. This point was accepted by some figures in the tradition, of course, but Dewey's emphasis of this feature allows him to criticize transcendent ends as incompatible with it. "In a strict sense, an end-in-view is a means in present action; present action is not a means to a remote end."[69] In this sense, Dewey can rightly call his position Instrumentalism[70] since ends are instrumental both as means to or motive for some present objective and, when achieved as consequences, as possible future means. Even general ends and notions of general value are conceived of as instrumental in a situation. The value is in relation to a situation and partly defined in terms of that situation. Growth of an organism or the value of health can be in a different direction in one situation than in another.

Dewey's Critique of Traditional Moral Theory

Dewey was critical of the main traditions in Western ethics based on his experimentalist refinement of the relation of means and ends and what he regarded as warranted attention to close examination of circumstances and consequences. Some of these points were covered in passing in the earlier sections, for example, that for the tradition in general, ends are singular, fixed, and bear only a coincidental relation to actual circumstances. Dewey traces this obsession on fixation to the Greeks, particularly Aristotle. He believed that the Aristotelian belief in fixed ends still pervades moral theory. Aristotle also originated the model in which the highest value, the *summum bonum*, is identified with ideals, which has created a "discrepancy" between existence and value, the ideal and the actual, theory and practice.[71] The result has been that most other consequences are ignored in favor of the single fixed end. This has created a "bias" in which deductive reasoning from certain fixed premises supports the idea of "complete determination and finality."[72]

Dewey's critique of fixed ends includes a notion widespread in the tradition of an end in itself, which is singular, fixed, and thus beyond any consideration of circumstance and consequence. The notion of an end in itself is utilized by a wide range of value theories, from idealist to naturalist and ancient to modern. Dewey believed that this notion provided the justification for the doctrine that the end justifies the means, as a single fixed end excludes consideration of other consequences and justifies any means used to achieve it. Dewey's critique of fixed, singular ends in themselves is also an implicit argument against actions done for their own sake, as a definition of virtue. Actions done for their own sake are ends in themselves, a notion Dewey finds contradictory. A further criticism is that fixed ends refer the significance of present activity to some end beyond it, which depreciates the value of the activity. *Summum bonum* theories in particular involve a "subordination of activity to a result outside itself."[73] Although this may seem like an argument for the value of a present activity done for its own sake, the latter actually is measured by a fixed end, which has no immediate relation to the activity itself. The value of the activity done "for its own sake" is, for Dewey, actually measured by an extrinsic standard, and thus not for its own sake, but for the sake of the standard. This model actually depreciates present activity for the sake of some future goal. But it is the present situation that should be our primary concern.[74]

Dewey applies these criticisms to specific ethical theories. Idealist theories provide a prominent example of the separation of fixed ends and actual conditions. Idealism locates value in a beyond of the supernatural; in the merely subjective, isolated from external conditions; or in an a priori realm that is removed from experience of the actual. The ideal as perfection by contrast with the actual has its origin in Platonic transcendence; its source is the gap between impossible goals and activity. Values based on actions are uncertain and insecure for this tradition, which thus posits a need for ideal values and essences. Idealism separates ideals from the actual world and the latter is treated "as a display of physical forces incapable of generating moral value."[75] Dewey argues that this theory involves the separation of desire from thought, of value from knowledge, and of intrinsic and instrumental value. The trouble with such an approach is that it does "not get beyond the stage of fancy of something agreeable and desirable, based upon an emotional wish."[76] An intelligent study of conditions is requisite to yield a practical and working end.

Dewey particularly criticizes what he regards as the a priori approach of Kantian idealism in these respects. The notion of the end in itself as a fixed, a priori, transcendent standard by which desires are judged exemplifies the problem with morals for Dewey. He argues that this approach involves the corollary that no control of desires, and thus of value properties is possible.

This corollary is scored on the ground that "valuations are constant phenomena of human behavior" and that humans can develop good habits, can correct these and that the corrections are made in the light of such valuations."[77] The a priori approach separates to an extreme degree natural and transcendent value. But Dewey argues that no a priori hierarchy of value can be helpful in every situation, with its unique features. For Kant, it is primarily motives that justify acts, an ethic of intent for which the single fixed end of a good intent justifies all. Dewey regards the emphasis on intent as, in effect, a subjectivization of consequences: a state of internal feeling is the sole end. Dewey also thinks that the ethic of intent constitutes an attempt to try to evade responsibility for consequences.[78] On the level of the practical, Kant's emphasis on subjective willing ignores actual means to ends relations, in the sense of cause and effect, by separating practical will from nature. Dewey accuses Kant of naively arguing that "if the right end is pointed to them, all that is required in order to bring about the right act is will or wish. . . ."[79] Dewey also rejects the idea that one can "get results without intelligent control of means," that is, the use of naturally based cause and effect relations. Will is a "cause of consequences," not dissevered from them. Kant increases the gap between morals and experience, something that rightly ought to be attenuated.

Despite his own consequentialism, Dewey also criticizes the main consequentialist theories, Utilitarianism and the Interest theory. Utilitarianism is seen as a step in the right direction as compared with idealism in that its end is not transcendent and it argues for "the natural, empirical character of moral judgements and beliefs." However, it is not consequentialist enough in a sense, as its theory of value, hedonism, is based on a single fixed end, which ignores all other consequences, particularly those connected directly with the problematic situation and its resolution. In effect, Utilitarianism is not experimental enough. Like other fixed-end morals, such as Kantianism, Utilitarianism denies the relevance of time and of change to morals: time and change are irrelevant for a fixed end.[80] Thus the peculiarities of a novel situation are left out of the calculation of ends. Further, Dewey argues that the Utilitarians followed a "false psychology," based on the separation of the mind from nature, and that the dualism skewered the ethics based on it. Dewey argues that the end is not pleasure but the means of "removing obstructions to an ongoing unified system of activities."[81] Deliberation is not for the sake of calculating advantage; pleasure is more like a culmination of experience in an ongoing process than an end point in either sense: goal or consequence. In sum, Dewey agrees with other critics of Utilitarianism that it "degrades morals."

The Interest theory is criticized on the same grounds as the Utilitarian and idealist theories, namely, that it posits a fixed end as the only goal of activity. The Interest theory ignores the peculiarities of the situation, proper attention to means, other consequences, and the many other factors requisite for deliberation and evaluation of the problematic situation.[82] Perry and Prall assume that desire and interest are unproblematic, a position Dewey challenges. "Desire and interest are not given ready-made at the outset . . . original data or premises."[83] Desire is not a given, a fixed end, but arises in time and in a situation of lack, a problem for which a solution is desired. Desire emerges within a complex whole of a problematic situation, in which a desire for some resolution is emergent, not foundational, and is constituted by or in terms of the ongoing activities of the agent.

Summary and Conclusion

"Instrumentalism" is slightly misleading as a characterization of Dewey's theory of value, since Dewey does not deny intrinsic value. Dewey's point, however, which is in accord with much of the tradition, is that ends can subsequently be used instrumentally as means. Dewey simply emphasized this point and used it as a critical notion to attenuate the absoluteness of the distinction. The distinction of ends-in-view from consequences allows Dewey to point out that the former are a stage in a practical process, that is, are instrumental to results. Activities undertaken in view of an end are also means, but are the locus of value, since Dewey has an essentially active theory of value.[84] Further, means can have a value of their own, which, as a quality of them, is 'intrinsic'; this formulation is in accord with the standard meaning of intrinsic ("inherent" in C. I. Lewis). Thus "intrinsic" does not distinguish ends as a quality or attribute, although other elements may mark the distinction.

By attenuating the distinction of ends and means, Dewey argued convincingly for the instrumental role of ends. Ends, principles, and past experiences are instrumental in the unique value situation. More, Dewey revalued means as worthy of greater consideration, an extension of value, since means are not considered subordinate in value or of inferior worth in his analysis. Another extension of value is involved in the existence of a plurality of ends instead of a single fixed end. Finally, Dewey took consequentialism seriously, and emphasized both the role of consequences in modifying future activity, their didactic role, and also the regulative role they ought to have over activity, their ethical role. Ends are reincorporated into practice as ends-in-view and results in Dewey's critique of transcendent, fixed ends.

Another feature of Dewey's analysis is that ends are rational. This may seem strange because Dewey is a critic of classical rationalism. However, for the tradition ends are often "intuited," in keeping with Aristotle's foundationalism, in which not everything can be demonstrated.[85] Mill for example, argues for the intuition of ultimate ends, that is, pleasure. Thus ultimate ends are not subject to deliberation or evaluation for the tradition; critical evaluation is not possible if ends are fixed. Dewey argues, on the contrary, that ends are not fixed but situational, and thus subject to critical evaluation and deliberation. "Desires," simple, are not values, for they have not been critically evaluated;[86] prizings are critically evaluated desires. The connection of ends-in-view to the situation can be demonstrated, and must be for any action to be rational. Thus natural ends are rational, and reason is natural. In this continuity of means and ends and ends becoming means there is a practical movement in time[87] in which ends are naturalized. Ends are not timeless but within time, like other natural processes. Holistic justification, in which all the elements of deliberation are considered in relation to actual circumstances, makes the rationality of ends part of the process. I will return to this last point in the next chapter.

Dewey's analysis is radical in that value is separated from teleology. Again, this may seem strange in view of his Instrumental theory of value. However, ends-in-view are not the locus of value and ends as consequences require further evaluation. Value may attach to means, to objects, to activities, and still other bearers that are not immediately involved in the means to end relation. They are potential instruments, but not always actual instruments. Again, as intrinsic or instrumental value is a property of what has it, it is analytically separable from activity involving means to ends. Intrinsic value is not an end in itself, for Dewey denies the notion of ends in themselves as "self-contradictory," but does not deny intrinsic value. In this separation, Dewey breaks with the Aristotelian tradition, for which the "end or good" are equivalent. Because justification is holistic, the question of intrinsic value is less important as there are other elements by which the best action under the circumstances is evaluated. The question of the contemplative character of values is perhaps less important than the distinction of valuing and value: the role of good in activity as an instrument for resolving a problematic situation. The action or means is part of the very process of determining what is prized.

The good as instrumental is also within an overall process. Activities bring about new objects and so are instrumental to their existence. "Value propositions . . . exist whenever things are appraised as to their suitability and serviceability as means . . . but are about things to be brought into existence."[88] The agency of bringing about is human activity as a means; this function is provided by habit. Potential means must be organized to be effective in bringing about; this organization is the function of habits. Dewey argues that the "means or effective conditions of the realization of a purpose" are not independent of "established habit."[89] Habits are the means closest at hand in a situation and most within our power. In a process, we work backward from the end as mediate to the immediate, our own store of experience in like situations and our abilities embodied in habits. We also work with existing conditions considered instrumentally as a means, not simply as a problem. Physical events, although they do not intrinsically possess any "instrumental or fulfilling character," can be used as causes to effect some end in "organic action."[90] In the latter case, a natural but inorganic process becomes part of a purposive process.

Ends for Dewey are in nature. They are not timeless, transcendent, or ulterior to a natural process. "In fact, ends are ends-in-view or aims. They arise out of natural effects or consequences...."[91] Values are in nature and thus part of nature: there is no denial of "natural goods;" on the contrary, Dewey explicitly affirms "natural goods." Nature is again invested with value, both because human valuing is within nature and because the instrumental value of nature is the model for all value in some respect.[92] Thus the problem of the intrinsic value of nature does not arise as a special problem for Dewey's Instrumentalism. All valuing is within nature and valuing is ultimately instrumental. Treating nature as instrumental is valuing it in some respect because any end can later be used as a means. But humans are not an exception to this model, for they too are within natural processes.

Contrary to Katz's reading, intrinsic value is not denied. Dewey's "critique" of intrinsic value is more of a critique of the subjective turn in intrinsic value in the modern age: the assignment of intrinsic value exclusively to a subject, for which all of nature is merely an instrument. Subjective intrinsic value is undermined by Dewey's arguments against "mentalistic psychology" and the corresponding theories of value. The charge that the values of pragmatism are based on human "desires and interests" is equally untenable, since Dewey makes a sustained critique of the interest theory and uncritical desire. His critique of the conative theory is mainly of a subjectivistic interpretation of desire and interest, particularly that of Prall. The target is a subject removed from nature and thus of subjective values as removed from nature. Dewey explicitly argues against any "private, introspective view of the field of values."[93] His value theory overcomes the subjectivity of relational theories that absolutize intrinsic value in a subject and thus are the root cause of the devaluation of nature. Far from being "relativistic," Dewey's attempt to bring scientific methods to bear in morals is an attempt to get past the relativism of extreme subjectivistic theories by providing a scientific warrant for ethics. Dewey does not uphold an individualistic view of value or morals, as has also been charged; morals are primarily social and value connected to activity within a set of relations, including relations to a social environment.

The removal of the subject and the values of the subject from the natural world of passionless mechanism is undermined by Dewey's analysis. If the value relation of means to ends is like that of cause and effect, the locus of value is in the action used as a means or the object brought into existence as an end. The object is objectively, because demonstrably, of value as a cause bringing about certain results. If it is brought into existence as an end it is objective qua object or objective. In such cases the object must have value in the instrumental sense of tending to produce effects. But if all valuing is instrumental, then all valuing is objective in this respect. An object cannot be judged valuable—the value judgment in which a tool is judged valuable—unless the judgment refers to some property of the object. Otherwise the judgment and language are incoherent. The point is that ends are not removed from nature as subjective; ends can later be means and thus instrumental in a nonsubjective way. Subjective value is the expression of subjective states only: emotions, desires, feelings, willing, interests. These have no direct relation to what is valued. For Dewey such states may become part of the whole of value, but are not values as such.

Dewey has solved the "problem" of the subjectivity of values, in that instrumental value belongs to the character of the object in its function as an instrument in a circumstance. This instrumentality is part of a whole in which the organism adjusts more or less well to the natural environment: the direction of change. Dewey has also overcome the subjectivization of intrinsic value in relational theories, which caused the devaluation of nature: the location of absolute intrinsic value in a subject. *If a good or an end can later be used as a means in the world, it cannot be subjective.* However, to characterize Dewey's theory of value as "objective" would be misleading. An objective theory is defined in relation to a "subjective" theory, in which a subject is opposed to an object—in other words a Cartesian universe. Dewey's philosophy is the attempt to dissolve such dualisms and the problems that arise with respect to them, for example, the value of nature. Value is primarily connected with activity for Dewey, rather than the object. The object is brought about by activity as the objective. Its value is derivative, not in terms of worth but as event. Dewey is a naturalist and his treatment of human activity is naturalistic. Intrinsic value is naturalized by Dewey in his denial that it is fixed, transcendent, or outside natural processes. End value is naturalized, a natural model for the value of nature.

4

DEWEY'S (MORAL) HOLISM

Dewey believed that changing conditions in the modern world called for a new approach to morality.[1] His project was to direct moral thought away from an individualistic basis. The exclusive concentration on individuals isolated humans from each other,[2] to speak nothing of nature. Morals have an affect on the individual of course, reflecting social forces, or what he called the social environment. Morality originally reflected social folkways and customs, but when these come into question, as they often have in the modern world, morals must be treated critically. It is the task of philosophy to supplement customs with rational ends.[3]

Dewey is critical of the separation of theory and practice, of knowing and doing by most of the philosophical tradition,[4] as well as the separation of morals from the rest of practical activity.

> Instead of being extended to cover all forms of action by means of which all the values of life are extended and rendered more secure, including the diffusion of the fine arts and the cultivation of taste ... and all activities which are concerned with rendering human relationships more significant and worthy, the meaning of 'practical' is limited to matters of ease, comfort, riches, bodily security ... things which in their isolation from other goods can only lay claim to restricted and narrow value.[5]

Practice should rightly cover all activity; practice should be coextensive with morals. Morals are concerned with "all activity into which alternative possibilities enter," that is, where "a difference between better and worse arises." Dewey argues that this includes "potentially ... every act" within the scope of morals insofar as better and worse alternatives for action present themselves, that is, almost always.[6] The need for morals originates in the conflict of different ends, rights, and duties, and the office of morals is to reflect on and help resolve such conflicts over better and worse alternatives. Morality is not a special kind of action in this view, as its scope may take in all of action. Moral theory is an inquiry with the objective of expanding what makes life worthwhile, the good, throughout all human relationships. It is meliorist in the attempt to improve life through expanding the good.[7]

Dewey calls his method in moral theory "experimental." "It implies that reflective morality demands observation of particular situations, rather than fixed adherence to a priori principles . . . for trying different measures so that their effects may be capable of observation and of comparison."[8] Close analysis of situations is emphasized over fixed principles in this method, a point Dewey will emphasize again and again.[9] Trial and error are also used to determine effects or results, such that these can be evaluated. In short, the method involves tests for moral notions similar to and derivative from those in science, and involving the treatment of theories as hypotheses, experimental activities, and close observation of consequences. Dewey views such tests as both practically worthwhile and as a guard against stifling dogma, which precludes inquiry into moral questions.[10] Moral principles are not rejected, however; they are used as tools to help deal with the situation. The correct role of principles, rules, and other moral maxims is instrumental for either understanding the situation better or helping resolve a problem perceived in it.[11] Legitimate principles embody past experience; the appeal to extratemporal, immutable principles and ends are "without support."

Elements in the Problematic Value Situation

Dewey analyzes the concrete situation in which valuations arise in some detail. The elements in this analysis will be treated distinctly below.[12] However, their essential continuity and interconnections should be kept in mind. It should not be inferred from the analysis of elements of a valuing situation that the elements are somehow isolated from one another. The latter would be precisely the opposite of Dewey's analysis because he frequently stresses both the continuity and the interrelations of these elements.

The first element is a problematic situation itself. Normally, activity can take place based on the routine of habit and occasional impulse. Habits may reflect training, character, and established social conventions. However, problems may arise to upset this routine, some "trouble in the existing situation." Both valuation and moral reflection arise in a problematic situation, in which the problem must be resolved and organic equilibrium restored.[13] Until a disturbing situation arises, there is "no need, no desire, no valuation." Such disturbances cause a temporary unease, but in the long run may have a beneficial effect. "In fact, situations into which change and the unexpected enter are a challenge to intelligence to create new principles."[14] Disturbances in any routine in the form of problematic situations create the opportunity for growth, the primary end. By changing the conditions through activity as a means to some valued end, they may also provide new connections and meanings, further constituents in valuation. Thus Dewey can speak in one passage of the "good of the situation." The situation is not limited to that faced by an isolated individual, but includes social conditions as well, the social situation in which the individual must operate. These may also be problematic and require novel solutions. Dewey calls for remaking "social conditions so that they will . . . support fuller and more enduring values." What should be emphasized is that the problematic situation does not simply involve a subject but outer conditions and a social milieu.

The uniqueness of the circumstance is the starting point for appraisal of any value situation. Failure to study such situations closely will invariably end with bad results. Dewey, however, always qualifies the role of any one element of a problematic situation, and the situation itself is no exception. The situation "never completely dominates" due to the presence of habit; there is a fund of experience individuals can draw on so that they are not overwhelmed by any situation. Of course, routines that seem nonmoral in isolation can "derive moral significance from the ends to which they lead." Thus moral considerations can be "present" even in habitual actions.[15]

The next element in a value situation is desire, which may include interest, liking, and the other synonyms familiar from Perry, Prall, and the interest theory for a conative theory of value. Dewey submits the interest theory to a sustained critique without entirely jettisoning it: desire is incorporated as one element in valuation.[16] Desire is created from or out of a problematic situation, that is, it arises in a concrete context. "We change character from worse to better only by changing conditions. . . . The stimulation of desire and effort is one preliminary in the change of surroundings."[17] The problematic situation creates a desire to resolve a problem, a perceived lack, that generates the formulation of a need relative to the situation.[18] Just as success is tied to specific circumstances, so is the projected end: "in fact we envisage the good in specific terms that are relative to existing needs."[19] The object needed or desired is projected as good, in that it will resolve the problematic situation. Such a solution "performs the function of resolution of a problem for the sake of which it is adopted. . . ." Since valuation is connected primarily with activity, which involves action as a means to an end, desire can help in the formulation of such an end of activity by providing a possible object as an end-in-view. Thus it may generate activity that moves in a new direction away from the problematic situation toward its resolution.[20]

Dewey speaks of the "emergence" of desire and whatever will satisfy it as well as the "value-property" in an object, or other bearer that does satisfy desire.[21] The end in view is, as I noted earlier, the object or objective that hopefully will resolve the problematic situation, in terms of desire.[22] Desire cannot be separated from ends-in-view almost by definition, since a desire is a desire for some end that will cause the desire to cease. Ends-in-view are ideas about what will resolve the problematic situation constituted along with and in terms of desire: the objective. Objects are identified with goals, a distinct element in the value situation. Ends-in-view are constituted as working ends for further activity but should not be confused with consequences.[23] They are hypothetical ideas, which should resolve the problematic situation. The actual outcome may not correspond to the hoped for one; the end-in-view is an idea for action, but may turn out badly. However, such ends-in-view are not one unreflective desire among others that has been evaluated more worthy. Dewey argues that valuation occurs only through "a transformation of a prior impulse or routine habit"[24] Desire in the sense of impulse is transformed into the desirable through deliberation on means and ends. Thus desire simple, in the sense of impulse, or even routine habit, is not valuation. Desires must be constituted in terms of the situation.

It is important to stress that ends-in-view are not results, but ideas or images of what will alleviate the present problematic situation. They are constituted as "ideational" ends-in-view in relation to a warranted desire because it is believed that they will resolve the problem favorably. Thus in the value situation as a whole, they are means-ends, that is, the end-in-view is a means in the sense of a cause of the "activities by which actual results are produced."[25] Although achieving the end-in-view may solve the problem, constituting it is a means in the process as a whole, that is, a means to initiating activity that will bring it about. Thus it is an end-means, an end that is intricately connected with means. This is crucial to Dewey's instrumental theory of value, for it makes of situational ends, ends-in-view, a means, or instrument. Dewey argues that the end-in-view is warranted to the degree that it is "formed in terms of these operative conditions."

Evaluation of desires is important both in consideration of suitability for the specific circumstances of the problem, and in terms of the connection to the whole process. The distinction is characterized by the use of 'desirable' by contrast with what is simply desired. The desirable emerges from reflection on and evaluation of what is desired in terms of the problematic situation and the overall process.[26] Both elements are required for valuation, because "desirable" alone without an object would be blind, but is a condition of both striving or activity and evaluation of ends-in-view. What is desired is a matter of fact, not of value, even in the sense of value-facts. It is not reflective of the method of experiment or of reflection on desire in terms of overall goals. "All growth in maturity consists in not immediately giving way to such tendencies ...," that is, to impulsive desires but "remaking them" in terms of consequences and evaluating them. Giving in to impulses, or unreflective desires, is a mark of immaturity, of a lack of character. Desires alone are not enough to decide a course of action.[27] A context is needed to differentiate desires and to compare them to the problematic situation in order to evaluate different values by comparison with one another.[28] Evaluation of desires as better or worse results in a judgment of what is desirable.

The relation of evaluation to value is that value often results from evaluation of the best course of action, but the latter may in turn reflect past values. Thus a process is created in which evaluation and value can grow in terms of one another: they are "complementary." Valuations—that is, desires and the corresponding end-in-view—can be both compared and evaluated. Dewey believes that it is clear in certain situations that some desires, actions, and outcomes may be better than others, the content of evaluation. More "the evaluation may modify further direct acts of prizing."[29] Evaluation is a critical activity distinct from valuation, then, although it is applied to each of the elements of valuation. This may involve comparison of existing valuations with past ones enabling them to be "reevaluated" in terms of testable evidence gathered from previous activity. The valued objective is determined by appraisal of concrete means, tying valuation and deliberation together critically. Finally, valuations are appraised in terms of results, both anticipated and actual.

Deliberation on the situation issues in a choice, which involves judgment between the competing alternatives. Choice or decision between competing valuations, means, and consequent outcomes is the next step after evaluation of them. It is prior to action.[30] Choices should be decided in accord with reason or follow the course that has been evaluated best under the circumstances. Dewey rejects the idea that a rational choice is one that follows a fixed standard; rationality is more grounded in choosing the best for the present situation. It is notable that both desires and ends are chosen, not simply means: "chosen desires" are distinguished from impulses by having been shaped by the process that has been outlined above. Dewey argues that a decision should be considered hypothetical and tentative, in conformity with the experimental method, "until the anticipated . . . consequences . . . have been squared with actual consequences."[31]

Another element in a value situation is the role of intelligence. Intelligence has a crucial double role in deliberation and the formation of ends.[32] "Intelligence is concerned with foreseeing the future so that action may have order and direction." Along with imagination, intelligence is involved in the study of conditions, articulating ends or goals, in connecting them with means and in anticipating probable consequences of action. Without intelligence as an instrument, the unique conditions of the problematic situation could not be analyzed, distinguished, and compared. Further, hypotheses as means of resolving the problematic situation are formulated by intelligence, based in part on past experience. "There is present an intellectual factor—a factor of inquiry—whenever there is valuation, for the end-in-view is formed and projected as that which, if acted upon, will supply the existing need or lack and resolve the existing conflict." The role of intelligence is to articulate this objective, formulate it from the fund of experience, connect it with suitable means, and so on, based on an inquiry into the actual situation and judgments as to the suitability of ends. Intelligence is not "cold"; it is indispensable in mediating desire and the object of desire. Constituting an objective also requires imagining what will satisfy a duly recognized desire. In sum, values are "identical with goods that are the fruit of intelligently directed activity." Intelligence and imagination are indispensable tools for resolving problematic situations by giving activity a positive direction.[33]

In terms of the problematic situation, the results that may be brought about, the end-in-view, may require care, that is, activity to preserve what is prized. This is a consideration in bringing it about: is it worth the care it will require to preserve or maintain it? A new house may generate a desire on one level, but will require maintenance. The care it will require is a consideration in the decision to make or purchase one. However, if an immediate object is considered good, its preservation may also be judged good.[34] The care required is one consideration, not the only one.

What is the relation of satisfaction of desire to value? Dewey explicitly states that he agrees with the Utilitarians that the moral good, and indeed every good, "consists in a satisfaction of the forces of human nature, in welfare, happiness."[35] The problem with the Utilitarian view is that it is too future oriented and thus tends to delay present satisfaction. The point here is that satisfaction of desire can be good. Intrinsic values in the form of consummations of a process of activity are another consideration. Humans value such consummations and will pursue them. Enhancing such values may be a factor in evaluating alternative courses of action. However, they are not the only consideration, nor the foundation of the process, a point that will be covered shortly.

Dewey states that "the problem of valuation in general as well as in particular cases concerns things that sustain to one another the relation of means-ends." Consideration of means is not only another element in the value situation, but is central because Dewey's modification or transformation of it is crucial in understanding his theory of value. What should be stressed in this context is that consideration of means no less than ends, and as inseparably connected with ends, is an important element in valuation. As L. D. Willard has noted, means and ends form a continuum, whose value is as a whole.[36] It is important in the first place as the condition of the end, of resolving the problem. "The needful or required is that which is *existentially necessary* if an end-in-view is to be brought into actual existence." Thus there is a necessary connection between means and ends: means are required for ends as their cause and thus must be considered in formulating ends, a point Dewey stresses in his critique of autonomous ends. Another point is that means have to do with facts, namely, what is existentially required. This places means in principle within the purview of the scientific or experimental method and constitutes an implicit critique of noncognitivism: means are knowable and can be publicly compared. Theories, principles, and rules are not ruled out however: they are intellectual means that can be utilized as appropriate. The role of theory is in investigating or inquiring into the "things sustaining the relation of ends-means." The results of such an investigation can be brought to bear on the formation of ends, due to the necessary relation of means and ends. This involves moral deliberation, the process of considering all the factors of a problematic situation. In one sense, Dewey's ethics is above all a critique of the tradition of moral deliberation.

Dewey argues that the separation of theory from practice in much of the tradition has not tended to render the values conducive to a good life secure. The connection to overall values is that action as means contributes or is instrumental to sustaining life. An overall or "inclusive" value is "maintained in adverse circumstances," that is, despite them, and it must endure through such changes, over immediate circumstances and the ends connected with them. Overall good is not the direct end of desire, which would be contingent on immediate circumstances. It belongs to the self apart from circumstances as a state of one's character.[37] There is also the instrumental role of ideals, which, properly utilized, can be valuable tools for analyzing the problematic situation. Dewey identifies ideals with the highest good.[38] Consideration of both overall values and ideals is requisite in moral situations. This could take the form of whether a proposed activity is conducive to the growth of the organism as a life. The activity must be compatible with life taken as a whole and life always involves growth.

The final element of the situation I will consider in this section is its resolution or outcome, the achievement of the goal, the result or consequence. Since consequences are included in the original definition of pragmatism, it is to be expected that they will play some important role in resolution of the problematic situation and evaluation of the solution. This is indeed the case; ends-in-view are judged less in terms of their intrinsic value than in terms of their consequences in resolving the immediate situation. Consequences are ends of action and so ends of a type, but as I previously noted, Dewey distinguishes ends-in-view or goals from ends as results. The latter is "existential" rather than an idea and is the ground of the value of the former. "Valuation of ends-in-view is tested by consequences that actually ensue."[39] If the goal is not achieved, something about the proposed resolution is defective. Means, as necessarily connected with ends, are also judged by consequences, by efficacy in actually improving the situation.

It might seem as though Dewey was offering a consequentialist test of value as it were, that is, that value should be judged in terms of successful consequences. However, this would be misleading, as the outcome is only one element in the situation. Consequences are indeed important in evaluation of the end-means, but the activity also has connection to the overall processes of life and growth. Thus consequentialism is qualified by positive direction of growth or at least the long-term consequences have priority over immediate ones.[40] The consequences themselves can be evaluated from the perspective of solutions to the problematic situation, overall values, and activities. "Positive attainment, actual enrichment of meaning and powers opens new vistas and sets new tasks, creates new aims and stimulates new efforts."[41] The activity has a place in the life as a whole with its many different kinds of activities. A successful outcome involves the furtherance of this harmonic organization by the organism. "It is to resolve entanglements in existing activity, restore continuity, recover harmony, utilize loose impulse and redirect habit."[42] Further, the consequence is not a final end, but the end of a determinate activity, which may lead to others.

In this section I have outlined the elements that Dewey identifies in the problematic value situation. These include particular problematic circumstances, desire and its satisfaction, the construction of ends-in-view, the role of intelligence, evaluation and choice, the use of means and caring, overall and inclusive goods, and the actual outcome or consequences. These elements, it must be stressed, are united as one activity with a specific context and result. However, they may sometimes be temporally separated as stages in a process.

Dewey's Consequentialism

Just as one would expect from a pragmatist, Dewey is primarily a consequentialist in ethics in the narrow sense of obligations, that is, moral duties are determined and judged in terms of consequences.[43] Ends as consequences are the justification for means and obligation is defined in terms of such ends. Dewey endorses this connection of ends, in the sense of "actual consequences," and that ends provide the "warrant for means employed" as a "correct position."[44] The means to consequences relation is identical with the cause to effect relation; in this Dewey agrees with Aristotle and much of the tradition. However, this is not the end of the matter, for there are important qualifications to be made in the case of Dewey's consequentialism. Since morality is primarily social for Dewey, moral acts are concerned with social consequences or the social context of consequences, not egoistic ones. The consequences for society create or entail obligations or right. Second, morals also include self-development or self-realization, which has social consequences also, but is primarily concerned with development of *self*. For Dewey, of course, the self does not develop in isolation from society, but reflects and is dependent on a supportive social framework and environment.[45] Nevertheless, this element of morality is distinct from social consequences in some respects. Another point is that the present meaning of the act is of equal or greater weight than remote future consequences. One might say that situational consequences outweigh transcendent ones, but this is a qualification of the problematic of moral consequentialism as traditionally understood, for example, in Mill.[46] Finally, Dewey's consequentialism is not foundational, in the sense that intrinsic value provides the entire warrant for obligation. Thus Dewey has transformed "consequentialism" and provided a new model in which consequences are one important factor in moral justification, but not foundational.

Dewey's consequentialism is not conceived as a series of discrete acts since there are always relations to larger factors. Dewey argues that "error comes into theories when the moral goods are separated from their consequences. . . ."[47] Dewey rejects the separation of morals and valued consequences, as when the ends justify the means. Further, morals are instrumental in connecting the immediate problematic situation to larger, overall goals, in the process of moral deliberation. Their regulative function ensures some connection to growth, some larger meaning. Growth is not a fixed end point, but an ongoing process. "Morality is a continuing process, not a fixed achievement. Morals means growth of conduct in meaning . . . that kind of expansion in meaning which is consequent upon observations of the conditions and outcome of conduct."[48]

Dewey's approach to consequentialism constitutes a rejection of mere calculation of future advantage, as in Utilitarianism. As future consequences are difficult to predict exactly, or to control, Dewey regards moral judgments as hypothetical or "experimental" and "subject to revision" if they prove to be erroneous. "The good, satisfaction, 'end' of growth of present action in shades and scope of meaning is the only good within our control, and the only one, accordingly, for which responsibility exists."[49] Morals connect present activity to this ongoing process as a present meaning. Consequences for growth as a process involve immediate or present consequences taken as part of a larger whole. There is no gap between present means and future benefits, or between present activity and future consequences.

Since consequences are not determined in relation to a fixed end, the question arises: which consequences are to be regulative? How are consequences to be gauged? Dewey would answer in terms of superior solutions to problematic situations, which mark a positive direction of change. Since Dewey rejects the notion that any single consequence is paramount, picking out one end, the basis of traditional morals in the form of a *summum bonum*, is arbitrary. Many consequences must be considered.

The positive direction of change, which is a mark of growth, brings the act into relation with an ongoing process; the consequences for present growth are an important factor. Growth is a character of life, so the consequences for life are an overall consideration. Dewey argues that the question of the ultimate warrant or source of authority for morals cannot be answered, but in another sense "the authority is that of life."[50] Clearly, consequences detrimental to life, one's own or others', would ultimately end growth, the possibility of improvement, and so on—in other words, all goods.

There are other elements as well in relating an action to consequences, including social consequences, that is, consequences for society and social relations, as well as for betterment of the world and oneself. Thus Dewey's consequentialism involves calculation of the social consequences of a considered action.[51] Moral rules are instrumental, but social consequences are part of the calculation of ends-in-view. Dewey's form of consequentialism, like Prall's, includes morals in the form of social impact as an element in considering consequences. Morals are not merely instrumental, for the social consequences of an act are among the factors to be considered.

Dewey is aware that there is a connection between character and growth in the form of self-development. It would be misleading to view character solely in terms of social approbation. Still another moral consequence to be considered in any evaluation is the consequence for self and growth of self. To grow is to develop; a positive direction of change is a development located in an individual self. Dewey argues that this is fundamental: moral criteria attach more value to "what men and women are capable of becoming" than actual attainments; to "possibilities, not possessions." Development means, "what they can grow into."[52] Dewey cautions, of course, against setting up self-realization as a transcendent ideal or a single fixed end. Potentialities of development never stop, and Dewey thinks that we are always potentially able to improve. In a sense, Dewey's is an extreme form of self-development, for development and growth never cease, and even provide a test of the value of social institutions, which are decisive conditions of character formation. "The test of their value is the extent to which they educate every individual into the full stature of his possibility."[53] Social obligation is reciprocal in that just as the individual must take social consequences into account in decisions, society owes members equal opportunity for full development of potential.[54]

The richness of Dewey's pluralistic version of consequentialism should be apparent from this brief summary. By rejecting a single fixed end, Dewey can incorporate consequences for life, growth, self-development and character, and society while sticking to the solution to a problematic situation. Behavior is defined normatively as character, not a mere stimulus and response. But conduct takes place in a social and natural setting or environment. The life, growth, and development of the organism is as much a consequence of moral concern as consequences for others. All these factors must be considered in evaluating decisions in any problematic situation. Dewey has opened up the potentialities of consequentialism by breaking with the notion of fixed ends and foundational grounding in intrinsic value.

The Relation of Value and Obligation

Moral justification for Dewey is not foundational but holistic. There is no appeal to a timeless, static, fixed, "given" value or end. Indeed, means may be justified as much by circumstances and results as by their relation to an end. Ends are not denied either in the sense of goals or in the sense of consequences. However, the justification of ends is neither their intrinsic value nor a *summum bonum*, but a whole circle of features and elements brought to bear on the circumstances to resolve a problematic situation. The situation and consequences are perhaps central. Right is defined in relation to good, that is, consequentially. In moral deliberation consequences are one factor among others and the basis of right. Ethics concerns both the right and the good for Dewey, in accord with the tradition. The right is not reducible to the good, however, nor is Dewey a consequentialist in the foundational sense of utilitarianism.[55] Moral activities are incorporated into the overall activity of the organism by intelligence, trial and error, and so on.

Dewey is holistic and no one element is more important than the whole web of relations in moral deliberation. Dewey is careful to relate these to one another in a mutually reinforcing way. Thus circumstances are not the ultimate test, for their resolution is in terms of a positive direction of change, which includes both values, inclusive, overall goods, and reference to consequences. Results are justified by conformity to projected ends, which form a kind of test or standard for them. The end-in-view is not some ultimate standard for it must be in relation to the circumstances. In sum, each of the elements in the problematic situation has meaningful connections to the others and thus is not ultimate. Justification follows an organic model in which a whole of the problematic situation is analyzed in terms of its parts as elements, and must include relations and connections.

"Life as a whole" and life as an "ongoing system of activities"[56] are holistic formulas for integrating values. Holism is present in the rejection of foundationalism in which moral justification is based solely on a relation to individual subjective states or "ends in themselves." Dewey has attempted to overcome such dualisms as that of subject and object, act and character. There are pluralities of goods in relation to one another, which are factors in all moral deliberation. This relation of the many factors is like a relation of parts to a whole in an organism.[57] Because all factors together must be considered, including intrinsic value and consequences, moral justification cannot be reduced to an ultimate foundation. Again, Dewey's consequentialism involves the relation of right and good, but these are not the only factors in moral deliberation. The system of relations and meanings includes all the relevant elements of deliberation in the problematic situation.

By providing a holistic view, Dewey can argue against "metaphysical individualism" in ethics that it is anti-social.[58] Individual bearers and activities are primary, but the web of relations required for activity also has its place and value. This is consistent with Dewey's rejection of foundationalism in ethics, the culmination justifying all of the rest. On the contrary, many factors and relations are involved in problematic situations and each of these has its own role to play. He emphasizes the many "connections and transactions" required for valuation as opposed to the single relation of subject and object in relational theories. Further, general notions of value have instrumental value in a situation. These are arrived at by induction from individual cases of value, but still have a distinctive value as tools for solving problems.[59] Dewey is aware that a value bearer must be individual in the primary sense. However, the individual is in a web of relations that connect it in vital ways to larger wholes and specific kinds. Thus he notes the existence of behaviors like reproduction, which do not benefit the individual directly but the species as a whole: the good of the species or good for the species. The value relations may be more general than those attached to just the bearer. There is also a notion of species value in the notion of natural kinds. There is the recognition that the value realm must extend beyond the individual for the individual bearer to survive. The bearer of value is not always the sole locus of value.

Dewey's critique of foundationalism in ethics is also a critique of the ethics of perfection of character, which is now known by the unfortunate rubric "virtue ethics." For Aristotle, character is grounded in the natural fixed ends defined by human nature as a "state of being." Dewey introduces a new theory of character as active habits defined in relation to the overall process of growth as *change*. As Pappas has noted,[60] this undermines the dualism of act-centered views (the "ethics of doing") and character centered views (the "ethics of being") by incorporating growth and development, consideration of consequences, and other elements into a larger whole, where all the elements are considered. Character is revealed in action, that is, in habits of action.

Summary and Conclusion

Dewey saw that foundational models were based on a separation of theory and practice in Greek philosophy. Practice and its impact on the world, including its impact on the environment and nature, was not important in this model. Thus environmental impact was not a factor in moral decisions, since ends that were beyond the process justified it. This led to the separation of the theoretical subject from the world in Cartesian metaphysics and thus subjectivization of value. Rather than try to reestablish naturalistic foundationalism, which was the root of the problem, Dewey attempted to provide an alternative model in which intrinsic value is one but not the only factor in the larger web of relations. Moral considerability is based on a whole web of interrelated factors. Dewey's holistic moral justification complements the holistic approach to the environment called for by both Callicott and Rolston in different ways. The web of relations in nature is a superior model for an organism that is a product of and within nature.

Dewey accepts many ideas of traditional ethics, including the universal scope or application of morals, the strong incorporation of values, the relation of means and ends in some form, and the important role of rationality. However, Dewey's criticisms of traditional ethics have the ironic result of actually expanding the role of morals. For Dewey is an acute critic of the narrowing of ethics that has been the trend in modern thought. Morality grew out of the natural goods of life such as art and science, but is also a part of them, a holistic relation of mutual reinforcement. Dewey calls for "doing away with" the traditional distinction between moral goods, such as the virtues, and the other goods of life. Morals as social permeate the other goods insofar as they are social products or reflect social conditions. This expansion of the role of morals lends itself to and supports the expansion of moral considerability, which is called for by environmental ethicists of all varieties.

The major modifications introduced by Dewey are his acute and careful analysis of the elements of the problematic situation and their interrelation. Further, Dewey contributed to moral philosophy by the emphasis on the uniqueness of the concrete situation as the most important factor in the resolving the situation as problematic and his location of primary value in activity, not subjective states or objects. His emphasis on the mediating role of intelligence in deliberation and evaluation and the introduction of the experimental methods of science into moral deliberation was a new approach to classic issues. His views can be considered a naturalistic critique of the subjective, inward turn taken by modern ethics. His criticism of the separation of ends from means is more consistent with the use of certain ends as means in the tradition.[61] Finally, Dewey took consequentialism seriously, and emphasized both the role of consequences in modifying future activity, their didactic role; and also the regulative role they ought to have over activity, their ethical role. Ends are reincorporated into practice as ends-in-view and results in Dewey's critique of transcendent, fixed ends.

Eric Katz has charged that pragmatism cannot serve as an environmental ethic because it denies intrinsic value and is anthropocentric.[62] I would respond that Katz's interpretation cannot apply to Dewey in several respects. First, Dewey, and for that matter the other pragmatists,[63] do not deny intrinsic value. Intrinsic value is not rejected, but is made subject to evaluation and some connection with the actual circumstances. What is rejected is the foundational model for ethics in which morals are based on intrinsic value. Katz's own arguments that intrinsic value has been overemphasized by the environmental literature could have been lifted from Dewey. Dewey provides an alternative model to that based on intrinsic value in which intrinsic value is part of an ongoing natural process in an instrumental relation to the organic whole.

The process includes intrinsic value, but is not based on it. On the contrary, intrinsic value depends on a certain direction of the process. Dewey's holistic treatment brings intrinsic value within the natural, and reincorporates the human subject into the natural environment.

Dewey substitutes an active theory of both human psychology and value. Desire and interest are reinterpreted as conation, an active striving or behavior of an organism. Dewey's empiricism is of a behavioral sort, which is common to all organisms. End value is naturalized: a natural model for value in and of nature. Intrinsic value also is naturalized in Dewey's denial that it is fixed, beyond natural processes, and thus transcendent or outside nature. Growth as an organic process and overall end combines value in a natural event. Morality is indeed human, but human nature is within an environment, which demands moral consideration as necessary to human survival.

The method of intelligence involves a whole circle of justification: a new, anti-foundational model.[64] Dewey has moved from a foundational model based on intrinsic value[65] to activity as a process, incorporating many elements in various relations. Dewey is an organic holist, although the organic analogy can be overdrawn. The emphasis on circumstances describes the situation of the organism in the environment. His value theory incorporates conative elements and the means-end relation into a larger, contextual whole of the ongoing actions of an organism, "that bring them into systematic relations with one another."[66] The means-end relation is conceived in terms of a more holistic continuum.

5

DEWEY'S ETHICS AS A BASIS FOR ENVIRONMENTAL ISSUES

Dewey was environmentally conscious before the issues of the environment received that wider attention that they have in more recent times. He was acutely aware that humans, despite their advances, live in an environment that has limits to abuse. His philosophy could be used as the basis for an environmental ethic with very little modification, since it already places humans in nature, avoids foundational models based on a detached subject, which make an environmental philosophy difficult to justify, and argues for responsibilities to the environment. Such a change in awareness reflects changing conditions, which constitute a new challenge to human intelligence to "change habits."

Dewey argued in his work that it is the task of philosophy to replace customs with rational ends. The question is whether norms can be derived from Dewey's corpus given that he did not conceive the office of a critical morality to issue specific rules and precepts. Further, with his holistic approach, (intrinsic) value and even moral standing do not necessarily involve corresponding obligations, a point that anticipates such environmental writers as Callicott and Rolston.[1] The repudiation of foundationalism in the relation of intrinsic value to moral standing is also a repudiation of the foundational model for ethics, that is, of obligation as grounded in intrinsic value. Can specific obligations to nature be derived from Dewey for use in environmental ethics? Is Dewey's Instrumentalism adequate for extension as an environmental ethics? In this chapter, I will try to make a case for such norms. This is justified given the previous efforts at extending classical ethical philosophies to environmental issues by environmental philosophers, for example, Callicott's use of Hume and Singer's of the Utilitarians. It is also a response to the challenge laid down by radical environmentalists as to whether a new, environmental ethic is needed.

I propose to use Dewey's holistic approach to moral situations for specific environmental problems and issues. I will begin with a short summary of Dewey's holistic method.[2] The method will then be applied to specific obligations to the environment, including the environment taken as a whole, to other species, to future generations, to individuals, landscapes, and rare habitats. Finally I will consider some recent criticisms of Dewey in the environmental literature, and argue that these are based on misunderstandings of Dewey.

Holistic Justification

The hidden minor of environmental ethics, moral considerability, is derived from a consequentialist ethic, in which duty is a consequence of the good.[3] The Utilitarian model of Singer, the first of the environmental ethicists to propose the value of the nonhuman as an issue, has been surreptitiously adopted by many of the figures in environmental ethics. In this model, moral considerability is based on intrinsic value and moral obligation on moral considerability. The question is whether this is the only way to justify ethics. Dewey argues that it is not. Dewey has a different model of the relation of value to morality than that of many of the radical environmentalists, such as Callicott.[4] Specifically, Dewey rejects "fundamental" or "foundational" grounding of morals in intrinsic value. Morality is not based on intrinsic value, although intrinsic value is part of naturalistic processes. Thus Dewey would reject the model of basing the moral standing of nature on its intrinsic value.[5] The advantage of his model is that, unlike intrinsic value theories, he can account for the value of the whole without trying to also argue for the "rights" of individual animals, an impossible project given predation and the health of the ecosystem as a whole. He is not caught in Rolston's bind of attempting to defend both the intrinsic value of individual organisms and also the importance of "capturing" this intrinsic value in predation.

Dewey saw that foundational models were based on a separation of theory and practice in Greek philosophy. Practice and its impact on the world, including its impact on the environment and nature, was not important in this model. Thus environmental impact was not a factor in moral decisions, since ends that were beyond the process justified it. This led to the separation of the theoretical subject from the world in Cartesian metaphysics and thus subjectivization of value. Rather than try to reestablish naturalistic foundationalism, which was the root of the problem, Dewey attempted to provide an alternative model in which intrinsic value is one but not the only factor in the larger web of relations. Dewey's theory is not an alternative foundationalism in which obligation is based on instrumental value instead of intrinsic value. Rather, both intrinsic and instrumental value are morally considerable elements among others. Moral considerability is based on a whole web of interrelated factors. Dewey is holistic and no one element is more important than the whole web of relations in moral deliberation.[6] Dewey in effect has proposed a novel thesis of moral considerability, in which the many factors that must be considered in any problematic situation are all relevant to moral decisions. Since they must all be considered, they are all morally considerable: all the elements considered in a problematic situation are within the scope of morals. Above all, the relations of all these factors in time and space, as an ongoing process in the web of relations, must be considered.

For this reason, Dewey's ethics is more suitable as a basis for environmental ethics. The scope of morals is wide for Dewey and includes all acts. Moral considerability, however, is not grounded in acts alone, but involves consideration of a number of elements. These elements form their own web of relations in deliberation just as natural elements form a web of relations in an environment. An environment is a web of relations more than a mere collection of individuals. The elements of moral considerability include, first, a problematic situation, a perceived condition that is considered to be disturbing, bad, defective, in need of change, and so on. Second, desires should be evaluated for desirability in relation to the situation. This is also the place for choosing the "end-in-view" or imagined solution. Third, intelligence is required to resolve any problematic situation, including reference to similar situations in the past and the use of scientific knowledge. A positive direction of change is fourth, whether a proposed act will actually improve the problematic situation in a melioristic direction and restore harmony among conflicting ends including valued consummations of experience, that is, intrinsic values as factors. Another consideration is that the proposed solution will generate new activities, which require caring for what is prized. Sixth, is consideration of means, whether actions are suitable, moral, and so on. The relation to overall or inclusive values, life, and growth is a factor regulative of immediate good. Finally, whether or not the outcome of the proposed action actually has the consequences that were aimed for is the test of the proposed solution, and another element to consider. Above all, the relations of all these factors in time and space, as an ongoing process in the web of relations, must be considered.

Obligations are what is required for carefully evaluated ends in the sense of consequences, that is, ends that have been intelligently considered and incorporate all the factors of moral deliberation, not just intrinsic value. Intrinsic value is included in moral consideration, but is not the ultimate basis for all the other elements. Moral obligation, then, is consequentialist, but only in a qualified sense: if the consequences have been carefully considered in terms of all the elements in relation.

I would argue that obligations to the environment can be justified based on Dewey's holism. The argument is that since the environment is required for humans, and what is required is obligatory,[7] that obligations to preserve the environment are obligatory. The environment is not a luxury but a necessity for humans. Degradation of the environment considered as a whole, and essential elements in it such as air, water, and soil, are almost a paradigm case of a problematic situation. The destruction of the environment is self-destructive on the part of humans. This clearly constitutes a problematic situation, since all human values, including "subjective values," require the maintenance of the environment as a whole in good health, caring for it. Pollution of the environment as a whole is problematic for all life and in particular human life. This is the problematic situation of modern humans with their present level of technological achievement. Increases in population also affect the environment in problematic ways; raising enough food to feed increased populations has negative impacts on the wild in terms of destruction of and disappearance of habitats. Soils have been degraded and diminished, air made unbreathable and water, polluted. Narrow interests, justified in terms of economic demand alone, aggravate the problematic situation, leading in a negative direction of change. The whole situation must be looked at, what Dewey called a "broad, just, sympathetic survey of situations."[8] Consideration of a situation in terms of narrow issues, for example, economic factors or the value of one single species, is strongly discouraged by this approach.

Since a problematic situation has been identified with respect to the environment as a whole, a positive direction of change is requisite, which will result in improvement of this situation. Intelligence must be utilized to identify each aspect of the problematic situation so that the problems can be dealt with. The function of intelligence in this regard is the gathering of information, including past solutions, the theories that have been formulated by environmental sciences and other intellectual resources for use as a tool in determining the best course of action. Dewey also argued for the role of intelligence in critically evaluating desires.[9] Intelligence and evaluation have the role of mediating impulses, that is, deliberation about whether impulses, including desires, should be acted on. Evaluation of possible solutions for overall superiority is part of intelligent deliberation, and may include experimental possibilities that are novel solutions to novel situations, echoing scientific experimentalism. Trial and error may be among the possible options of intelligent action. The sources of pollution must be analyzed, with each factor contributing to soil degradation singled out; and motives for increasing population, such as family labor and security, determined. Specific habitats that are threatened or endangered should be mapped. The impact of current technology on the environment must be intelligently treated. Science, far from being a source of the problem in environmental affairs, can aid in the solution. This is indicated by the vanguard role certain scientists have played in calling attention to environmental problems in the first place.[10]

Consideration of means is also imperative. Just as blowing up a whole city with an atomic device as a means of getting rid of one old building would be considered insane, so are other inappropriate means. Thus the end of profits may not justify the means of cutting down an old growth forest, which provides a unique environment, a home to many species, preserves soil by channeling rainfall, and is an unparalleled source of consummatory experiences. Looking at the issue in terms of profits alone is unintelligent, arbitrary, and narrow. Evaluation of the means of dealing with and resolving the problematic situation posed by the environment as a whole is imperative and must be considered along with ends. Means that worsen the situation are clearly no better than continuing the present course. Contour plowing may save the soil, but this will have little impact on the problem posed by the environment if population increases result in continued expansion of acreage under cultivation. All the web of relations must be considered in evaluation of means, including human satisfactions and other values. Many means that are now in common use may in the future be ruled out if obligations to the environment are derived from consideration of this whole. Thus new means must be invented to deal with the changed situation.

The ends in the sense of results that are considered to deal with the problematic situation ought to lie in a positive direction of change. The change should solve the problem of the degradation of the environment as a whole, or at least move in that direction. The end-in-view of improving the environment as a whole must match results. If the methods tried do not achieve the results projected, they are ineffective and ought to be reevaluated. Improvement must be measurable and considerable. Hopefully such ends will increase human satisfactions in contemplation of the environment as a whole and specific, especial niches within it.

The results must also be compatible with the overall values of life and growth. Damage to the environment from global warming could at some point be catastrophic for human and other life. Such overall goods are an important factor in regulating more immediate ends, which might be harmful to growth, for example, dumping dangerous chemicals into the ground, which can threaten groundwater. Overall life and growth must be maintained in the environment, which may involve control of individual species or other elements, including soil and water. These overall ends are regulative over immediate ends: a basic factor in Dewey's critique of foundationalism in ethics.

Changes in the environment and in our activities must somehow be an improvement, representing a positive direction of change. Morals are concerned with better and worse, with bringing about improvement in a problematic situation. Cutting individual trees in an old growth forest that are diseased, and thus a threat to the others, may well be such an improvement. What must be kept in mind in such decisions is that the consequences of such actions may generate new conditions, which require new actions, if the environment is to be maintained and properly cared for.

Finally, the web of relations in any environment taken as a whole must be considered. Essential aspects of this must be identified, for example, the impact of grazers on plant populations, and the role of the latter in cleansing the air, channeling the water, and holding in the soil. If grazing of livestock negatively impacts an environment, overgrazing, then it may have to be cut back or terminated to allow the environment to recover. Without such recuperative measures the environment may be permanently altered such that no grazing will be possible. This would be a self-defeating outcome or consequence; it would also aggravate an already problematic situation. The consequences would be bad, a negative direction of change.

What is primarily at issue for environmental ethics are consequences for the environment and Dewey is strongly aware of the value of the environment to both the human organism and other organisms, and explicitly argues for taking the environment into consideration. The social and environmental consequences of any acts must be considered; the impact on the environment affects the society as well since the latter is in the former. Morals are not divorced from the environment, but within it. Morality is social for Dewey and the social consequences of an act are one important aspect of consideration of consequences. Morality is necessary for the social environment in which we live. The individual requires a social environment to live and thrive; in turn this requires a physical environment. The impact on the physical environment can be considered an impact on the social environment and thus as within the scope of moral consideration.

The issue then becomes whether a case can be made for the nonhuman having moral standing. If morals are primarily social for Dewey, how can morality include the environment? The answer is that the social is defined as a specialized environment within the larger biological and physical environment. If morals are social, then morality is concerned with the proper or optimum environment for humans. Both the social and the larger environment as the place for the social are included. Read in this way, Dewey is an environmental ethicist, for he refuses to separate the human from the environment, or the social environment from the biosphere. Morality is concerned with creating the optimum environment for human flourishing, growth, and life. Destruction of this environment would have bad consequences for the human organism and its particular social milieu. Since all acts are within the scope of morals, acts that involve environmental consequences are within moral purview: they should be considered in deliberation. Dewey has an ecocentric outlook and the environment must "have its say."

A second consequence that ought to be considered, then, is impact on the world, or the environment as such: the environmental consequences. Dewey notes that one consideration in moral decisions is what sort of world we want to make by our actions. What sort of world do we wish to have or live in? This practical problematic cannot be divorced from environmental ethics, since we live in the environment as organisms. "Bringing about" a better environment is imperative, given the norm of meliorism. Ignoring the environmental impact of pollution or global warming is self-destructive: it is for the worse, not the better. The definition of value in terms of "caring for" or preserving what is prized is required for the environment, thus an obligation or duty.

Finally, there is the impact on the individual, on the person. This is in keeping with the question, what sort of person do we wish to become? The consequences for our self-development, our character, and personal happiness are still another element in consideration of consequences of our activities.

Environmental Ethics

How do all these elements affect specific issues in environmental ethics? Are there obligations to the environment or biosphere as a whole based on Dewey's holism? Can obligations to species be derived? Do animals, plants, and the environment have moral standing? What of animal liberation: are there obligations to individual higher animals as sentient? Do we have obligations to future generations? What of landscapes, specific habitats, and other elements of the environment such as soil, air, and water?

Clearly, moral considerability is not confined to humans, given Dewey's account of human nature as intimately bound up with a web of environmental relations.[11] Destruction of this environment would have bad consequences for the human organism, which requires it for its life processes. This reading might seem to contradict texts in Dewey in which morals seem to be a human affair, for example, "morality is largely concerned with controlling human nature."[12] However, not only is human nature within nature as a whole, but separating human nature from nature has adverse consequences for human nature itself.[13] "A morals based on study of human nature . . . would find the facts of man continuous with the rest of nature and would thereby ally ethics with physics and biology."[14] Human morals are within a larger context of nature in general and related to humans as biological creatures in a physical world. Life and mind are "highly complex" characters of events, that is, within the natural. Dewey argues that morals cut off from human nature is "bound to be negative"[15] and calls for appealing to something in human nature in order to realize ideals. Humans may be the sole agents of morals for Dewey, but morals cannot be cut off from nature because humans are within nature.

Good also attaches to the environment itself as the necessary condition of life and activities of the organism. The environment has instrumental value, but of a distinct sort, for it is not a random or coincidental instrument but vital and necessary. Habits and other activities *require* the environment in which to operate. "Habits involve the support of environing conditions...."[16] The environment is not only needed for activity but supportively, as an instrument. Many habits would be impossible without utilizing elements and tools taken from the environment, for example, such crafts as woodworking. Such skills actually incorporate elements of the environment into the activity.[17]

Life also requires an environment to exist and flourish. As far as is known, no life can exist outside a suitable environment, and the latter includes such elements as air or water, a nutritional source, a place to inhabit, and the like. These are instrumental to life, but in a special sense for they are both condition and part of the organism. The organism is "in and of" a specialized environment suitable to it as a species and without which it could not survive. Thus the environment is not merely an instrument, but in a sense a part of the organism. "Life, for example, involves the habit of eating, which in turn involves a unification of organism and nature."[18] The environment is not simply "out there" confronting the detached subject, but part of the constitution of the (human) organism.

These considerations are the basis for an environmental ethic, in the specific sense of moral obligation to the environment as a whole, on several grounds. First, the environment is required for organic life and its activity. What is "required" in order to perform or continue "a course of activity" is right, according to Dewey.[19] Right is often expressed with moral principles and rules. As the environment is required for the maintenance of any and all activity, it comes under moral consideration *a fortiori* as obligatory. Dewey explicitly recognizes a responsibility to the environment. Any "rational basis" for morals "must begin with recognizing that functions and habits are ways of using and incorporating the environment *in which the latter has its say as surely as the former.*"[20] Second, the environment enters into activity and organic life in intimate ways: they are in a sense unified. Insofar as activity and organic life have value, the environment shares in this value: they are inseparable. Dewey argues that the complete separation of consciousness from the physical is untenable in view of "its specifiable connections with organic conditions and . . . the intimate, unbroken connection of organic with inorganic events."[21] Further, the environment has instrumental value for the organism and its activities—indeed it is an essential instrumentality for them. The consequences of destruction of this environment would be harmful to the organism in multiple ways, direct and indirect. An environmental ethic is requisite based on negative consequences to the organism of harm to an essential instrument. Further, the environment as a whole might be seen as a direct value bearer whose destruction is detrimental to ongoing processes. Finally, animals and plants might be seen as bearing value since they are behavioral organisms in the relevant respects.

Humans require nature and must adapt to it. "Man needs the earth in order to walk, the sea to swim or sail, the air to fly. Of necessity he acts within the world and in order to be, he must in some measure adapt himself as one part of nature to other parts."[22] Dewey thought that the environment was not as flexible or plastic as, say, the materials of poetry. Humans have to adapt to the needs of the environment since organic life, and the minds that organism make possible, cannot be separated from and require an environment. What is required is that the good of the environment as a whole is considered in any alteration of it. This principle provides a source of norms to conserve the environment as a whole, as distinct from what is in it. Because it is distinct from what is in it, the norm is not anthropocentric, despite its origin as an anthropocentric concern. Dewey describes this as "fidelity" to the nature to which we belong. Fidelity to nature as such and as a whole is requisite, apart from human desires.[23] This virtue requires caring for what is prized: looking out for the environment as a whole in our activities. In order to preserve what is valued as part of ourselves, it must be prized and cared for.

More, a "positive direction of change" must be measured by a more adequate adjustment to the environment by the organism. This could be used as the basis for a norm requiring a better adjustment to the environment by humans. Adaptation to the environment is required for human survival; destruction of it is self-destructive for the species as a whole. A better treatment of the environment would be required as the content and focus of this adjustment. Human desires would require appraisal in terms of their adequacy with respect to adjustment to the environment. Desires that would impact the environment in destructive ways, for example, the desire to use all-terrain vehicles in fragile desert landscapes, would be ruled out. Such desires require reconsideration in light of the norm of adequate adjustment to the environment of the human organism. Successful coping requires functioning well in an environment, not simply achieving consequences. Good is not limited to consequences, but involves many other factors, and consideration of other forms of good, particularly functioning well in the environment.

Dewey does not base moral considerability on intrinsic value.[24] However, since all acts are within the scope of morals for Dewey, activities that involve the environment or elements within it are morally considerable. All value has some moral standing in the sense that its value should be considered before it is wantonly destroyed. Unless its destruction would actually aid in a solution to a problematic situation, it is not a positive direction of change. The goal of activity is improvement and wanton destruction cannot be justified on this basis. Nor can it be justified in terms of single fixed ends, which override all other considerations, for example, economic profits. The end cannot justify immoral means for Dewey, and destruction of the environment would have bad consequences for all of life. Further, the consequences of destroying it may be deleterious, if only because it cannot then be used instrumentally in the future. This would argue against the destruction of "useless" species, since their potential future use would be foreclosed.

In many cases, obligations to the environment override specific ends of individuals because their social impact would be harmful. The environment as a whole has moral claims, which are of higher significance than ill-conceived individual ends. This principle extends to essential elements of the environment such as air, water, soils, and essential species (e.g., bees). Short-term interests do not provide sufficient warrant for destruction of the biosphere and the corresponding impact on the society. The biosphere is the condition of and required for the social environment, which is the condition of individual growth and thriving. The environment as a whole with its web of relations has greater moral standing than individual ends, since it is required for the latter. Our obligation to it overrides the standing of harmful acts with merely short-term benefits. The environment is required for humans in a supreme sense since it is part of what we are. It is not merely morally considerable: we have obligations to it that may often overrule lesser goods.

A strong case can also be made from Dewey's ethics for moral obligations to preserve other species based on their essential role in specific environments, as an essential element of the whole. Predators, for example, are required to keep grazers in check, such that overgrazing does not occur. Overgrazing would not only result in the end of the value of the habitat for life but of its use by grazers: a self-defeating consequence. Overgrazing is one problematic situation in specific habitats. Upsetting the balance of species in a habitat is a more general one. In some cases, human intelligence may be of aid in setting things right, even though the problematic situation may have arisen due to human activity. Human impact can be studied to determine how it impacts particular vital species, which are defined as such due to their essential role in the whole and thus value for life overall. Means could be evaluated for efficacy in alleviating the problematic situation. Caring for threatened species vital to a habitat could be considered and undertaken as a project, which would have betterment of the situation as the end-in-view. The results could be checked against this standard to determine if the measures taken to preserve the species were effective. Ideally the consequence would be restoration of the species and an improved, healthier habitat.

An even stronger case could be made for the moral considerability of nonhuman species in that they share certain essential aspects of human nature that are morally relevant. These include life, growth, sentience, intelligence, and the struggle to survive. Dewey viewed preservation of life, or survival, as a species imperative. Preservation of life is an ongoing process that transcends individuals, since, as Dewey noted, parents of all species must sacrifice for the sake of their offspring, a behavior that does not benefit them directly. They are contributing to and thus often part of the web of life in time, from generation to generation. Growth is required for maintenance of a life as preservation of a species; this involves caring for such an individual life. "It is also an obvious empirical fact that animals are connected with each other in inclusive schemes of behavior by means of signaling acts . . . [such that] a joint action made possible by signaling occurs." Animals have the rudiments of a language or at least some communication and also what appears to be purposive behavior. Since Dewey has defined all of these characteristics as morally relevant, and in some cases the measure of value, their presence in nonhuman species would require moral consideration of them. Further, his project entails the revaluation of nature, that is, that value is a trait of nature. If animal life is experienced as valuable, then this value is a real trait of nature, not a mere projection.[25] The experience of value in nature, then, is morally considerable, including the experience of the value of living things. It is important to note, however, that the spirit of Dewey's whole philosophy is against extension from the human to the nonhuman in favor of the inclusion of the human within the natural. The arguments for extension of moral standing might strike Dewey as odd, since humans should never have been conceived as exceptions to natural processes.

Dewey also argued that social cooperation did not entail the submergence of the individual or what is unique to the individual. By extension, a norm of preservation of the unique could be derived: what is unique to the species. Another norm is equality in the form of equality of opportunity.[26] This norm could be extended to species: all species have equal opportunity. From such a norm a case for species preservation could be made. The species is a distinct natural kind and the good of the species helps preserves a distinct natural kind. Dewey recognizes such natural kinds as goods. There is a notion of value of kinds of value, for example, in the classification of "such natural goods as health, wealth, honor . . . friendship . . ." and moral goods.[27] A basis for articulating the value of species as natural kinds could be derived from this consideration because value may attach in the abstract to the species. Dewey favors pluralistic kinds or bearers of value over monistic value schemes. Species contribute to value pluralism insofar as their preservation conserves distinct goods of nature. It might be argued that this is to run together kinds and value, but, as organisms, species represent distinct kinds of value embodied in organisms as well. Consideration of the end of preserving the environment has equal standing with economic ends, given value pluralism.

A species is necessary for the health of the whole, that is, the web of relations that constitute the biosphere, the living environment. Biodiversity is one mark of a thriving environment and each species may have its place in contributing to the whole. There may be more specific roles that elements may play in the environment as well. For example, predators may be required to keep the ungulates in check. Otherwise, the grazers may overgraze and ruin a habitat. Such roles are consistent with his principle that ends may later be used as means, which is how predators use prey. Instrumentalist value theory is more suitable in describing the web of life in which, to use Rolston's terms, natural value is "captured" in predation and used instrumentally in the food chain. The use of ends as means provides better norms for environmental ethics than intrinsic value models tied to nominalism.

Dewey also argues that a more social conception of ethics does not mean uniformity, but rather is the true condition of diversity.[28] This principle could be used as justification of a norm of natural diversity in the web of life. If the whole is of value, the environment itself, then each of its elements also has value. To say we cannot determine the role that a species plays is unpragmatic: it is premature to rule out future discovery and possible instrumentality in both the environment and for human use.

This argument can be put in a non-anthropocentric form, however. As Norton has noted,[29] there is another, more radically ecological view. The value of species in a niche, the holistic system of relations that gives "niche value," is often, although not always, irreplaceable. Niche value is less anthropocentric because value is assigned not by analogy with human instrumental value, but in relation to an environmental whole. This argument is reinforced by the moral equality of all instrumental values as equally instruments. The good that is instrumental to one species is the moral equal of that required for another qua instrument. The value of the instrumental role it plays in the environment is equal to that of others. It should be noted again, however, that Dewey is not substituting a foundationalism of instrumental value for that of intrinsic value. Both are elements of moral decision. Preservation of species requires preservation of its habitat as well.[30] In many cases this should be obligatory and override short-term goals and ill-considered desires. I conclude that obligations to species are strongly implied by Dewey's work, although extension is required to derive such obligation. Morals are not confined to humans; since obligation extends to the environment, it extends to essential species within the environment. Bees are an example, since they are required for pollination of many plants, including most of the plants prized by humans. Diversity, pluralism of natural goods, and other norms argue for the preservation of species.

The value of individuals of a species could be treated as morally considerable but not obligatory. Not only is the view that all of life is sacred logistically impossible, it is ecologically unfeasible, as Callicott has convincingly argued. Predators require prey to survive and this requires that some lives of prey be sacrificed. More, if we allowed the survival of the AIDS virus it would be at the expense of human lives.[31] This reading is also consistent with the principle that ends may later be used as means—in other words, Dewey's Instrumental theory of values. Individual animals used as means may benefit the whole, both human and other animals of the same species. Individual living things are morally considerable; Dewey argues that they should not suffer unnecessarily. However, he endorses the use of animals in experiments to discover medical cures for humans.[32] Thus there is no obligation to nonhuman individuals. Least considerable would be nonliving things, with qualifications for landscapes, perhaps. An individual rock would have no value apart from its place in a larger whole, or as an instrument. Its value would depend on its place in a larger environment. In placing humans first, Dewey is not reverting to a surreptitious anthropocentrism. Even the most radical of the environmental ethicists make a moral exception for humans, that is, Callicott and Rolston.

What of obligations to future generations?[33] If we can avoid global warming in order to prevent worse situations for our descendants, should we? The impact of present activity upon future generations is also problematic. Intelligence has often foreseen consequences of present activity that are negative or bad, for example, the overuse of fossil fuels. Their use has created a problematic situation that is novel but whose consequences are not immediate but long-term. Means of creating usable energy, which do not result in the greenhouse effect and thus will not result in long-term alteration of the environment, are required. To continue to ignore the problems created by means of fossil fuels is to put the short-term good ahead of the overall good. This is not a positive direction of change, but will end in even worse consequences. It will also impact the environment as a whole—that is, all of life—in unforeseen ways, possibly harmful ones. The negative direction would include harm to elements of the environment essential for the human environment. Thus it will have harmful consequences for human society as a whole as well as individuals within it.

If we prize future generations—our descendents—and care for them, we are obliged to create conditions for their well-being. Thus there is a challenge to intelligence to come up with long-term solutions that will be fair to present and also future generations. A case can be made for such obligations based on Dewey's view of nature and its necessity for human survival. Because natural processes are necessary to our survival, we have an obligation to protect the process itself. The process is an ongoing one by nature, thus the preservation of it as ongoing is requisite: it must go on into the future. The advantage of Dewey's view of nature is that it does not consist primarily of objects but of events and thus has a time factor built in. Time as history is an element of the process as a whole; thus self-preservation requires preservation of nature as temporal, including future time. Caring for ongoing processes of life that are prized is required for any life, since the environment is required for life. Also, as I noted previously, preservation of the life of the species is a natural imperative for all species in the form of survival. This creates a strong consideration for future generations, since their survival is necessary to preserve the species. In turn, a natural obligation to the future is created. Dewey also argues for the creation of habits that will include a broad survey of consequences. As consequences are in the future, deliberation ought to take the future into consideration. This would provide a framework for obligations to future generations.

Such obligations also impact human populations in the form of imperatives to limit population growth. While preservation of species is requisite, production of the maximum number of individuals within it is neither required, feasible, nor desirable. On the contrary, such increases would exceed the carrying capacity of the planet. Limiting population growth is necessary both for the survival of the planet as a whole and that of future generations, since the latter is dependent on the former. Individuals who are not born cannot have rights, only those who are born. This separates the issues of obligations to future generations from birth control. Obligations are to individuals actually born in the future, not potentially so. Since the difference between the two is a matter of present choice, it comes within the sphere of morals.

The building of dams in the western United States in order to create reservoirs has often resulted in the destruction of scenic canyons, notably Hetch Hetchy in Yosemite National Park in California and Hoover Dam on the Nevada-Arizona border. The dam blocks the flow of water behind it, filling in the canyon. This is a problematic situation in which the scenic value of landscapes is in conflict with the value of stored water for agricultural and other uses. There is a conflict of ends here suggestive of a problematic situation. The conflicts of rights and values can be viewed in Deweyan terms as a kind of problematic situation, since there is a conflict of ends and means, a lack of harmony, and no clear, decisive resolution for unified action. Dewey might suggest that the situation be examined in all its many details and ramifications, with both scientific and public input, to obtain the maximum "sympathetic" survey of the situation. Dewey argued that economic activity comes within the scope of morals, so the economic factors behind building the dam are part of the decision process in resolving the situation. Is the use of water for agriculture worth endangering a local variety of fish in the Colorado River that exists nowhere else? Does the end justify the means? Are other feasible sites for a reservoir available that would not spoil the scenic beauty of a proposed site, or the irreplaceable drawings made by local indigenous peoples of the other site? This would be the place for intelligence and imagination in coming up with new plans of action, new means for achieving both ends, preservation of unique landscapes and fauna varieties, and also retaining precious water for human use. The consequences of a mistake might mean irreversible damage to unique sites; this may not be a situation that calls for trial and error. Such consequences must be taken into account along with the end of retaining water: means along with ends. Alternative sites may be available, which would not mar the unique, scenic beauty of one canyon or the irreplaceable art on the walls of another. Identifying and utilizing such sites would mark a positive direction of change, a betterment of the existing situation. Alternatively, a reservoir could be dug away from valued, prized landscapes. While this might increase costs, the last are only one factor under consideration.

The value of a unique species is morally considerable and obligations to it are strongly implied, as I argued in the last section. This is a strong competing claim against the need for water for economic development, which is not a present claim, but a future goal outside the present situation. The value of economic development as a fixed end, subordinating all the rest is unjustified in Dewey's view. It is only one element of the situation, with little present claim. The situation as a whole involves a plurality of values, esthetic, moral, and economic, as well as relations to overall values. The value of economic development must be weighed against the value of unique landscapes. Scenic landscapes may also bring culminations of experience that are highly satisfactory and contribute to happiness.[34] Both self-development and happiness are ends that are morally considerable. Unique landscapes may provide unique culminations of experience or be intrinsically valuable. While obligation is not grounded in intrinsic value, the latter is a factor in moral decisions. A holistic theory would take all the values involved in a conflict into account and weigh them.

I would argue that a strong case could be made using Dewey's approach for the moral considerability of scenic landscapes, such as the Grand Canyon. Dewey revalued nature in the sense that the experience of value is a trait of nature. If beauty is experienced as a trait of a landscape like the Grand Canyon, then this is an actual trait of the landscape. It is tied to the experience of the Grand Canyon, not every experience. The more unique a landscape, the more it is prized and the more humans are willing to care for it. Even littering might be wrong if it defaces the environment, as it would not mark a positive direction of change. The ultimate question for moral decisions is what kind of world we wish to make and live in. A purely economic world would involve a diminishment of value and of kinds of value, subordination of all value to economic value. The economic model resembles the model of grounding in intrinsic value that Dewey opposed and criticized. Unless such landscapes are protected, distinctive experiences that contribute to human life might be lost.

The same considerations apply to specific habitats, unique environments that may have little recreational or esthetic value but are important for life. Indeed, Dewey might argue that consideration of specific habitats and unique environments are the best approach. This habitat is essential, that forest is threatened, a unique wilderness or rare species are at risk and so on. Such an approach is dealing with specific, concrete problems. Wetlands are the prime example, although not the only one. Wetlands are crucial for wildlife and are often essential to species based in them. They are often the locale for feeding of birds and so essential to the food chain of species that do not dwell there as well. Wetlands in the United States are disappearing at an alarming rate; they are being eliminated for development of housing and agriculture. Other examples are the habitats of rare species, such as a plant that lives only in one locale. Transplantation usually does not work; the plant is where it is because it has adapted to that site and usually does not do well off of it. Elimination of either of these kinds of habitats is a problematic situation, since it diminishes unique goods, even if only instrumental goods. As with unique landscapes, alternatives to development of such unique habitats should be studied. Subordination of all other factors to the one end of economic development is imposition of one value at the expense of all the others. The overall value of life should be a check upon the rash, impulsive desire for short-term profits. Excessive destruction of wetlands might have negative consequences for all of life including human life.

The right to dispose of private property is within a social environment, in which such rights are recognized. Moral rules are necessary for any society to function, and society in turn requires an environment suitable for life. The consequences for society as a whole of the continued destruction of unique habitats may be so harmful that society must limit certain private rights for social self-protection. I would argue that a strong case can be made along such lines for protection of specific or unique habitats. Actions on private property that have social consequences, for example, poisoning groundwater by dumping toxic chemicals into the ground, are commonly recognized to come under public regulation, the law. Morality is primarily social for Dewey, although it also includes individual development and satisfaction, and the social consequences of actions on private property cannot be ignored.

Similarly, elements of the environment as a whole, such as air, water, and soil, ought to be protected. Requiring soil conservation, for example, may be a social obligation, a recognition of the rights of future generations, a case for which was made above. More, these constitute essential parts of the biosphere as a whole, required for all of life. An obligation not to pollute is required for the health of the whole.

The negative mode in which obligations are often expressed sends a different message than the more positive outlook Dewey wished to emphasize. In his call for a new moral outlook, Dewey also promotes the positive idea of the individual as taking part in the community, as sharing. Morality includes more than obligations to society and the environment, important as these are. Other elements of Dewey's moral philosophy are the consequences for the individual, personal consequences, which include both self-development and personal happiness, satisfaction, enjoyment, and the like. Self-development involves asking what sort of person we wish to become. Nature may bring lessons crucial for self-development, what Norton called "transformative values." The environment may also be required for certain activities integral to self-development, for example, hikes in the wilderness. Happiness may be a prized motive for activity. Landscapes and other prized environments may bring unique personal pleasures and satisfactions and also contribute to overall happiness. The environment taken as a whole is required for any self-development or personal happiness whatsoever. Thus there are obligations to it given the value of self-development and happiness.[35]

Because the community forms the social environment, which in turn is in the larger biological and physical environment, this could provide an incentive for conservation, namely, positive participation in the environment. Participation in the larger environment is required for individual life and growth, since the environment is "in and of" each individual. Conservation is caring for what is prized as a part of one's self in a very important sense, the environment in which we live. Such activities accord with the overall value of our own life as well, a long-term consideration. This is a more positive approach to conservation and improvement of the environment than negative prohibitions.

Dewey also notes the benefits to the individual of social cooperation. This principle provides another norm that can be used for the treatment of environmental issues, that is, benefits to the individual of treating the environment well. Since the individual is ultimately dependent on this environment, this benefit is obvious. I already noted that the environment must be taken into account in Dewey's view; this point is reinforced by its benefits to the individual. It is right or obligatory as required for any benefits to the individual whatsoever: the individual could not survive without the environment.

Summary and Conclusion

Dewey has provided an alternative, holistic model of morals, which includes both consequences and other elements involved in resolution of a problematic situation. Moral considerability is not based on intrinsic value, although intrinsic value is part of natural processes. His holistic model of moral considerability and justification provides a better model for environmental ethics, in which a larger number of factors are taken into account than just monistic notions of the *summum bonum*. However, this approach strengthens the case for the moral standing of the environment as a whole and essential elements within it. All acts of humans are within the scope of morals, and the consequences of harm to our own environment, in which humans live and of which they are part, must be considered in moral deliberation. The principles of improving the world and ourselves can be used as factors in environmental decisions. Certain moral obligations to the environment as a whole, to species and to future generations, can be derived by extension of Dewey's naturalism. A strong case can be made for the preservation of landscapes, unique habitats, and essential elements of the environment based on Dewey's moral writings and his view of nature. Obligations to individual living things, humans excepted, is not recognized, which is an advantage from the point of view of environmental ethics. Predation of individuals may be necessary for the improvement of the species, the life of other, predator species and the health of the environment as a whole. Because we are physically situated within our environment, we are obligated to take it into account in rational deliberation over our own future. Consequences for the environment include consequences for ourselves.

The connection of this holistic model to the web of relations in the environment is evident, since Dewey's thought was formulated with nature and the relation of the organism to the environment in mind. For this reason Dewey's ethics is more suitable as a basis for environmental ethics: an environment is a web of relations more than a mere collection of individuals. The web of relations in nature is a superior model for an organism that is a product of and within nature. What is primarily at issue for environmental ethics are consequences for the environment, and Dewey is not only strongly aware of the value of the environment to both the human organism and other organisms, but explicitly argues for taking the environment into consideration. The social and environmental consequences of any acts must be considered; the impact on the environment affects the society as well since the latter is in the former. Morals are not divorced from the environment, but within it. Morality is social for Dewey and the social consequences of an act are one important aspect of consideration of consequences. Morality is necessary for the social environment in which we live. The individual requires a social environment in order to live and thrive; in turn this requires a physical environment. The impact on the physical environment can be considered an impact on the social environment and thus as within the scope of moral consideration.

Dewey's reformulation of the problems of value is better than intrinsic value models in several respects. Dewey's model provides an alternative to intrinsic value theories based on an independent, autonomous subject. Dewey's holism avoids the problem of grounding environmental ethics in intrinsic value and then backtracking on either the issue of the intrinsic value of individuals or the relation of intrinsic value and moral obligation.[36] If all individuals in nature have intrinsic value, their use as food by other species, an instrumental relation, is problematic. The continuity of the means-end relation is a more suitable model for actual natural and environmental processes, and is used as such by Rolston. Dewey's model of the use of ends as means is echoed by the intrinsic value environmentalists in their notion of predators using prey, for example, Rolston's "capturing natural value." In the "Land ethic," Callicott argues that an end is not ultimate, but may become a means. Dewey stresses the interdependence of the organism with its environment, a model much closer to nature: he is an ethical naturalist. By treating moral obligation in relation to situations and consequences, Dewey more nearly approximates the thrust of environmental ethics as concern over the consequences of impact on the environment.

Dewey's philosophy is especially relevant to environmental ethics as a thoroughgoing naturalism. His holistic treatment brings intrinsic value within the natural, a treatment that anticipates Rolston's analysis of natural intrinsic value. His philosophy does not have to be reworked, if properly interpreted. It is already naturalistic, organic, and environmentally minded. His ethic is easily applied to nonhuman nature, since it is naturalistic in orientation.

With his concentration on action as the primary sense of value, valuing becomes objective, not subjective. Value is attributed to objects, but especially to objectives. The objective of action can be observed and thus predicted: objectivity is redefined and relocated. The object is only secondarily 'objective' as it has been brought about by the natural activity of organisms. Properly speaking, Dewey's value theory is neither subjective nor objective, although it includes subjective and objective value. Value is connected with activities and action: an active, naturalistic theory. Action is within the natural environment and part of natural processes. The process is not fixed; value is intrinsic to the process but not foundational to the process. Value is transferred to the public space and within the context of prediction. Thus action for the future can be evaluated objectively for environmental impact. We can "bring about" changes in our relation to the environment for better or worse, which are melioristic or harmful. Evaluated facts are a special class of facts, which have been deliberated on, chosen, and brought about. Valuing is taken out of the realm of the arbitrary and into the natural world. Carefully considered values can be the source of a change for the better in our relation to the environment, a positive direction of change.

Epilogue

Pragmatism and Environmental Ethics

The basic reason an environmental ethic is needed for our age is the threat to future human survival posed by the poisoning of the human environment itself by industrial and other processes. However, the massive destruction of habitats, whose root cause is human population growth, has evoked a radical response from environmentally minded philosophers. They have raised the question of the value of nonhuman life and whether treating animals and other living things as mere resources is justified. Just as slaves, women, and others were at one time given less moral consideration, so, they argue, nonhuman life has been treated merely instrumentally. An environmental ethic is needed to defend nonhuman life, the environment as a whole, and endangered species in particular. This ethic must be based on the intrinsic value of the nonhuman, since intrinsic value is the basis of moral obligation in ethics. Pragmatism was attacked by some of these figures for contributing to the problem with an instrumental theory of value, a subjective and anthropocentric approach to philosophy, and indifference to nature.

I have tried to defend Dewey against such charges in chapters two through five. Dewey is, of course, only one pragmatist. Can pragmatism as a whole be defended against such charges? While defending all the pragmatists against these charges would be beyond the scope of this book, some general comments can be made about the other major pragmatists. First, there are differences among the pragmatists over some of the issues raised by environmental ethics. To take intrinsic value, Dewey was the only pragmatist who was critical of this notion as it was handed down to him. As I have tried to show, he was more critical of the notion of grounding ethics in a fixed notion of intrinsic value than intrinsic value per se. But in any case, the more idealistic pragmatists defended intrinsic value. Second, the whole thrust of pragmatism was away from Cartesian subjectivity. Pragmatism does look to human consequences among others. But such consequences are in a context or environment that is not grounded in an isolated human subject. Pragmatism is not an anthropocentric view, since humans are placed back in a context that requires attention to the natural world. With the stress on action as an agency of knowledge and value, all the pragmatists shift attention from subjective perception of objects to interaction with the environment. Action is in an environment, not detached from it.

I would argue that not only does pragmatism escape the above criticisms but provides several fertile grounds for an environmental ethic. These include both environmental ethics based on intrinsic value (the standard approach) and alternative approaches. I will first argue that there is a strong notion of intrinsic value in the more idealist wing of the pragmatists and that it could be used as the basis of an environmental ethics. Second, since the Cartesian standpoint is what pragmatism is moving away from, the elements of Cartesianism must be shown to be foreign to pragmatism.[1] I will argue that pragmatism in general is primarily a critique of Cartesianism. Third, I will argue that certain elements of pragmatism could indeed be the basis of an environmental ethics. An environmental ethic can be extracted and defended based on the differing philosophies of the pragmatists other than Dewey. Finally, pragmatism even provides an example of an alternative framework for environmental ethics based on ethics itself. From such a framework ecological ethics could be extended such that it provides the basis for the subject, and not an extension from the subject.

The problem of the intrinsic value of the elements of the environment arose as an issue for philosophy because nature was judged instrumentally valuable by the preponderance of the modern tradition. The distinction of intrinsic and instrumental value generally corresponded to the distinction of subject and object, in a devaluation of the external world and nature. Even where the nonhuman was seen as directly valuable, as in scenery or landscapes, it was regarded as valuable to or for a subject, upholding the distinction of subjective value as intrinsically good and objective value as merely instrumental. The value of the external world was limited to the aesthetic and instrumental, and did not include nonhuman life. Further, value in the subject was tied to the affective or to desire by many philosophers, especially empiricists, separating value itself from cognitive status. So extreme was the subjectivization of value that the ontological status of value became an issue for modern philosophy. Is value a direct property of the object or only attributed to the object by a subject? Can any real value attach to the nonhuman at all?

By contrast with Dewey, the idealist wing of pragmatism upheld the strong form of intrinsic value in which it provides the rationale for instrumental value. Further, intrinsic value was judged objective, as belonging to objects themselves. Peirce maintained that there were three normative sciences, and that, of these, the science of aesthetics has priority, for it tells us what ought to be valued for itself.[2] This constitutes a challenge to the priority of the moral in Dewey and Lewis. However, Peirce was also a realist in metaphysics. What is valuable in itself is real, not merely ideal, and the aesthetic represents a real or objective general quality of the world. On this basis a justification for the value and moral considerability of beautiful landscapes could be constructed. It need not be morally obligatory to have some moral standing. Peirce also undercut the notion of subjective value with his arguments for community and its formation of the individual, a point that will be covered in more detail below. Values cannot be entirely subjective if mind reflects the community.[3]

C. I. Lewis gave still another account of intrinsic value. Lewis tied intrinsic value to the subject and so seemingly can be viewed as evidence for the subjective reading of pragmatic value theory, as in Weston and Katz. Intrinsic value is located in the subject and extrinsic or instrumental value in the object, a division similar to that of other modern, relational theories. For Lewis, intrinsic value plays a foundational role for instrumental value: "the notion of objective value in a thing which is incapable of conducing, either directly or indirectly, to any immediate satisfaction, would be an absurdity."[4] Not only is subjective intrinsic value upheld and foundational to instrumental value, it is given in perception.[5] In all these respects, and despite his claim of continuity, Lewis's theory of value breaks with Dewey's and represents a return to the subject. There is much more of an emphasis on the subject, perception, and experience in the traditional, empirical sense.

However, although Lewis did indeed argue for the subjectivity of intrinsic value, this is not the end of the matter. In the first place, there were also elements of continuity with Dewey. Like Dewey and the other pragmatists, he acknowledges the important role of action in cognition as well as a strong consequentialism in moral theory and a defense of cognitivism in value theory. Lewis was particularly concerned to show that values and morals were not irrational or relative, a stance that he took emotivism to foster.[6] Lewis viewed himself as neither a relativist nor an absolutist in value theory; he was in the middle between "tough-minded" emotivists and "tender-minded" idealists. Value may be ascribed to objects as conducing to satisfactions. Objective value is indeed instrumental, but it can guide action as predictive of a future state. It is not irrational as predictive: use of an object will generally result in satisfaction. Value is not "merely" subjective because it is inherent in an object.

Lewis worked out a persuasive view of the ontology of value in which various distinctions of value are located more precisely. In particular, Lewis argues that some objects must have "inherent" value in order to provide extrinsic value to a subject. For Lewis, objects can be inherently valuable: valuable on their own. He distinguishes, that is, inherent value as the property of an object that has value on its own from the intrinsic value of an experience for a subject.[7] Inherent value, like instrumental value, is extrinsic. It involves a subject in the experience of value. However, an artwork may be valued "for its own sake" by an experiencing subject. The value of the artwork is inherent, according to Lewis, as conducing to satisfactions simply in the beholding of the work. This solves the problem of the ontology of value in relational theories. Where is value in a relational theory? What is the bearer of value? It is not the subject, for a relation to an object is required and vice versa. If it is in the subject alone, it does not involve a relation, and value cannot attach to the object. A purely subjective view could not take pleasure in the object, or be interested in it, or have feelings for it, or it would devolve into a relation. The relational view in turn requires some notion of the inherent value of the object or it reverts to a purely subjective, solipsistic view. By arguing that an object may have inherent value, Lewis clearly locates the value in a relation in both the object and also in the subject, not in a nebulous relation that is in or a quality of neither. Although Lewis's theory is relational, it is a relational theory with greater support than subjectively based theories. His formulation provides a stronger basis for the inherent value of nature than in other relational theories (e.g., that of Callicott), a point to which I will return.

All extrinsic values, instrumental or inherent, are objective, and not dependent on experience. "The thing 'is what it is regardless of what anybody thinks or feels about it,' as truly in the case of value as in the case of any other objective character."[8] Thus the object can be a value bearer independently of the subject because it may conduce to valued states of the subject. This issue is closely related to the bearer of value. That is, a relational view requires some property or character of an object that will normally excite subjectively desired states. If value is located purely in a subject, then it does not involve a relation to an object, and the object cannot be said to have value. But then no object can be the cause of valued states of the subject in a relation. The power of an object to affect a subject is an objective property of the object. In effect, Lewis argues that a relational theory that does not attribute objective value to the object is incoherent as it does not locate any value in the object. This view is also compatible with Dewey's instrumentalism. The value of the object as an instrument is objective, for example, a tool's value. Indeed, Lewis argues for the instrumentality of inherent value.

Finally, the *summum bonum* for Lewis is life as a whole, "the good life represents the *summum bonum*."[9] Life taken as a whole constitutes a norm for judging individual experiences of subjective value. For some subjectively based theories, intrinsic value and the *summum bonum* are run together, for example, hedonism. Although there is a sense of judging individual pleasures in hedonism, life as a whole is viewed as good if it contains the maximum quantity of pleasure. Individual pleasures are ruled out only if they interfere with other pleasures. Lewis has a much stronger sense of *summum bonum* as attached to the whole apart from intrinsic value.[10] Extrinsic value is grounded in intrinsic value, value for an experiencing subject. But intrinsic value is in turn judged by its relation to the whole. It is not final or ultimate. "The final end by reference to which all values are to be appraised, is the end of some possible good life . . . to a life which would be found good in the living of it."[11] Satisfactions are less active than life as a whole, which requires the "goods of action."[12] In this respect Lewis's theory is closer to Dewey's, for it is more holistic, based more on action and less on intrinsic value. It also posits life as an ultimate value distinct from any intrinsic value of a single experience and regulative over the latter.

Lewis is not ultimately a subjectivist, despite the subjectivity of intrinsic value and the grounding of instrumental value in intrinsic value. As in Dewey, intrinsic value is present but not ultimate. His theory can be misleading since it starts, as with most of the pragmatists, from experience, and this is likely to be interpreted as meaning subjective experience.[13] Action is required for knowledge: it is not based merely on perception.[14] Holism of value is connected to life, not individual experiences. Lewis describes his theory as "naturalistic,"[15] for nature is in some sense a model for human nature. Human nature is examined within the framework of nature, as in Dewey, and as opposed to Moore. The question arises as to whether Lewis's is an anthropocentric naturalism or can include nonhuman nature in the scope of value. Value as a natural quality of objects may imply nature as a whole as the context of value relations and thus the condition of subjective valuing. It was previously shown that objective value can belong to objects and inherent value to objects valued for their own sake. This approach creates a strong basis for including natural objects of value within the scope of value. As life as a whole is the ultimate value or *summum bonum*, by which individual values are judged, the life of other creatures gains moral considerability. The inherent value of nature is contributory to valued internal states of humans. "Nature too presents us with objects having inherent value and worthy of the most sensitive appreciation."[16] Lewis's distinction of inherent and instrumental value could be used to support the value of nature. Nature could be experienced as an object or a totality of objects with value qualities. Certain of these have inherent value, for example, aesthetic wild objects such as landscapes. These are conducive to satisfactions in experience, but are experienced for their own sake, not instrumentally. Landscapes, the beauty of wild nature, contact with wild creatures, may all fall within the scope of value. Again, action and life take place in an environment, not detached from it. Thus the environment has a stronger claim on us than other extrinsic values, for it is required for life and action.

It was noted earlier that Lewis is a consequentialist in ethics.[17] However, he puts more emphasis on ethics as the condition of certain values.[18] There is thus a stronger sense that (intrinsic) value can generate obligation in Lewis than in Dewey. Imperatives are interpreted consequentially, not simply in terms of universality and consistency, as in Kant. What is morally right is the condition of action aiming at a specific end as a consequence. Lewis' theories could provide the theoretical underpinnings for obligations to the environment by extension, based on its inherent value and its necessity for human life. As consequences to the environment are the main moral problem for environmental ethics, Lewis's ethics, like Dewey's, would need very little adjustment to be made over into or extended as an environmental ethic. Because value as an attribute is cognitive and predictive for Lewis, degradation of the environment due to human predation is objectively measurable and can be controlled. As for animals, pleasure and pain are used to train them, which is evidence of sentience. If the value experience is basic and liking or satisfaction is evidence of subjectivity, then animals have such experiences on Lewis's account.[19] They should have moral standing based on such an analogy to intrinsic value in humans. Lewis hints that this is the case: that an animal ought to have moral standing "in the degree that it is capable of enjoyment and suffering. . . ."[20]

Lewis could be used as the basis for an environmental ethic by those who cannot conceive of value apart from a human valuer but reject the notion that it is based on a human valuer: value is an objective feature. Lewis's position is stronger than Callicott's in the sense of presenting a more well-thought-out relational theory. Value is also connected with action and is objective. His theory could be used to justify saving an endangered species or a threatened habitat apart from their utilitarian or aesthetic value: the satisfaction of having preserved them as such. The pragmatic question arises of whether we need to go beyond the inherent value of nature or the moral standing of the environment so long as nature is preserved and treated with respect. The distinction of intrinsic and inherent value may be a tempest in a teapot if both amount to preserving the environment.

Critics and even proponents have also accused pragmatism of being subjective and anthropocentric in orientation.[21] Katz went so far as to claim that pragmatism is supportive of an "egoistic" approach to the environment. This is the most far-fetched claim of any. The whole thrust of pragmatism is away from the isolated Cartesian subject. This is indicated most clearly in Peirce, the founder of pragmatism. "The very origin of the conception of reality shows that this conception essentially involves the notion of a community, without definite limits, and capable of a definite increase of knowledge."[22] Knowledge is not based on the subject or individual inquiry, but is a cooperative endeavor with communal norms for correct procedure. Peirce completely rejects the notion of a detached subject who can provide the basis for knowledge or a measure of reality. Reality is not based on the subject, and this includes the reality of nature, which is independent of any subject. Residues of the Cartesian subject are expunged by the notion of community, despite the misleading appeal to experience in many of the pragmatists. Community in ecological thinking is matched by the idea of community in Peirce's philosophy, which is arguably the historical basis for its prominence in philosophy.

Shaner and Duval argue that William James's radical empiricism has an "egocentric" element that would prevent the necessary union between self and the natural world.[23] Shaner and Duval trace the influences of certain American figures on Japanese philosophy, particularly that of Louis Agassiz and William James on Nishida Kitaro, an important and influential figure in Japanese philosophy.[24] Agassiz, an early "ecocentric figure" also influenced James. Shaner and Duval distinguish an ecocentric from an "egocentric" outlook according to the criteria of attachment to the subjectivity of experience. An "ecocentric" outlook requires pure experience in the sense of "developed awareness between self and world," a "sensitive interaction between ... persons and their natural environment" and a "feel" for ecocentric interaction.[25] By contrast with the separateness characteristic of the individualism of the ego, the ecocentric world view is "holistic" and "characterized by an aesthetic and perhaps a material oneness with all things."[26] The ecocentric view contains a "conception of community in which there is no vestige of a Platonic ontological hierarchy of existence." Community, holism, and ecocentrism contrast with the "egocentric" view characterized by individual egos that subject or subjugate the ecology as a means, as in Cartesianism. "Intimacy [with the environment] requires egolessness such that one's experience is not attached to an intentional frame of reference from subject to object."[27] Shaner and Duval claim both that Japanese philosophy embodies the ecocentric world view and that James is ultimately egocentric since he was unable to achieve the detachment from the ego necessary to be ecocentric.

The gist of their critique is that James could never conceive of experience apart from a subject: of "pure experience." James conceives of consciousness as attached to a self. "James was not able seriously to question the notion of self, ego or soul that frequently stands in the way of entering fully into an ecocentric, versus egocentric, world view."[28] They argue that 'soul' is retained from religion. But, they argue, experience of the world "without ego attachment is, of course, crucial for cultivating a mode of being-in-the-world that is intimate."[29] Attachment to an ego, to self, blocks ecocentric consciousness. By way of qualification, they note that James's notion of "pure experience" from his "Radical Empiricism" influenced Nishida. However, they argue that James's notion of the "stream of consciousness" does not abandon egocentrism, based on evidence of certain marginal comments by James. James is portrayed as essentially Cartesian, with experience and consciousness requiring a subjective ego that experiences. This attachment blocks true ecocentric consciousness, and thus is an inappropriate basis for an environmental ethic.

By way of criticisms of Shaner and Duval, it should be remarked that the whole movement of pragmatism is away from Cartesianism. While James is its least radical figure in this regard, holding to experience more than was comfortable to Peirce, its founder, his notion of "radical experience" and the "stream of consciousness" were as far removed from Cartesianism as one can go and still be remotely in the same tradition of a conscious subject. Indeed, as Shaner and Duval themselves note, Nishida was influenced by James's notion that the stream of consciousness is *prior* to the subject-object distinction in James. It is more basic than the distinction of subject and object within experience. James's project in part was to overcome this Cartesian dualism, or at least marginalize it, with an analysis of something more basic. The charge of subjectivity is contradictory. The subject does indeed exist in James's analysis, but is not basic. Thus an "ecocentric" mode of experience could be based on the "stream of consciousness" without bringing in the subject. Further, there is a distinction between acknowledging that experience requires an experiencer, consciousness a conscious agent, which is a logical requirement; and experience in reference to a subject. James escapes the latter; Shaner and Duval have not shown how the former is avoided by their "ecocentric" view. Experience may differ in the degree of ecological awareness; it cannot differ in kind if it is experiential: it logically requires a subject. Are they arguing for experience without an experiencer? Disembodied experience that is experienced by no one? Surely not; thus the reference to the subject is what is important to avoid: precisely James's project. Experience is primitive in James, prior to any distinction of subjective and objective, as Shaner and Duvall admit.[30] Further, there is a notion of community in James, while truth is not subjective but consequential. Finally, James argues that a "pluralistic universe" must include "a genuinely 'external' environment of some sort. . . ."[31] This is to speak nothing of the holism of most of the figures in pragmatism, and their explicit critique of Cartesianism. In sum, Shaner and Duval have not made their case against James and even less against pragmatism. On the contrary, pragmatism exemplifies the very ideas that Shaner and Duval wish to promote.

Experience never means subjective experience for the pragmatists. Lewis also rejects the idea of knowledge based on purely subjective experience. I have already documented how for Dewey the 'subject' is basically a reflection of the social environment in which the individual is raised and educated, a view echoed by Mead. The "subject" is formed by a social community, which is within a natural environment. In sum, none of the pragmatists starts from the subject and some are dubious of the whole notion of a subject. Further, the community as the source of the development of mind argues against the subjectivity of values. At most, value is derived from the relation of a subject to an object: it is never merely subjective.

Pragmatism is not particularly anthropocentric either.[32] To discuss humans is not to regard them as central. Humans are always placed in some context in pragmatism that involves a relation to something else: a situation, a community, an environment, nature. The rejection of the Cartesian standpoint is a rejection of any model of an autonomous human subject confronting nature, which could provide a ground for the subordination of nature for human exploitation. By denying the ground, pragmatism in general undercuts the rationale.

The lesson of pragmatism in this regard is that a metaphysic of experience, properly formulated, is compatible with a theory of value that is not entirely subjective, but allows for the inherent value of objects. An experientially based value theory can include inherent, objective, or other nonsubjective value. Experience is not automatically subjective, and this argument applies to value as well. More, the most impressive value theorists among the pragmatists, Dewey, Lewis, and Schiller, emphasize that value is in many instances to be analyzed with respect to action, in terms of the "objective," not perception, the "object." This is an important distinction, for action is public and cognitive, not private and irrational. They all agree that value attaches to life as the ultimate standard, not subjective standards like pleasure or feeling. The shift in focus removes value from the Cartesian subject and replaces action and value in a living environment in the world.

The third major issue is whether an environmental ethic can be derived from pragmatism. One specific question is whether there is in pragmatism an alternative way of looking at nonhuman nature? Again, there is in the idealist wing of pragmatism a distinctive view of the nonhuman, a non-instrumental view of nature as potential mind. I refer of course to Peirce's pan-psychism, in which materialism is seen as "impossible." "The one intelligible theory of the universe is that of objective idealism, that matter is effete mind, inveterate habits becoming physical laws."[33] Peirce revived the ancient animistic view of nature as alive and ensouled. A similar view is contained in the "Gaia" hypothesis of some environmentalists, in which the earth as a whole is conceived as alive and, by some, ensouled. This view of nature, which actually has a long pedigree, is certainly an alternative to the mechanistic view of Descartes or the materialist view. Peirce's pan-psychism underscores the error of trying to lump all the pragmatists together, as if their thought is a monolith or they never disagree.

Further, the pragmatists tend toward holism, not individualism or atomism.[34] The premise of many of the radical environmentalists is that intrinsic value is the basis of moral obligation. On this basis, for example, they argue that pragmatism cannot be the basis for an environmental ethic because it does not have a strong notion of intrinsic value. The question that arises is whether this is the only approach or the best approach to the environment.

We have seen how this model runs into problems in dealing with prey relations. There is the further problem of deriving obligation from value. The model is based on human intrinsic value and may not be suitable as an environmental ethic. This is especially apparent when considering the intrinsic value of a dynamic and temporally finite ecology in which forests and other habitats go through cycles, from domination by pioneering species to later domination by climax species. If pioneer species have intrinsic value, then are we required to save them at the expense of later species in the same habitat? Mass extinctions are another dynamic element, in which whole species die out, allowing new species, including humans, to arise. There is also the problem of the locus of value with intrinsic value theories. Value as a quality attaches as an aspect, but the value of the whole may be at issue in environmental ethics, especially in consideration of the biosphere and the esthetic value of some landscapes, for example, the Grand Canyon. The de-emphasis of intrinsic value in pragmatic theories in favor of holism may be more suitable for environmental ethics. The environment may consist in a web of relations rather than discrete individuals. The relation of parts to the whole may be more important than individuals and the web of relations as a whole more important than any part. This is the thrust of two of the more radical figures in contemporary environmental ethics, Callicott and Rolston. The environment or nature taken as a whole may be viewed as the source of nearly all our prizings, and thus as having special moral considerability.

The pragmatic view of nature may constitute a challenge to the "intrinsic value" model as the ultimate context of life, growth and action. We are "in" an environment for Dewey and the other pragmatists and thus the issue of its intrinsic value cannot even arise: it is an absolute necessity. The environment is beyond intrinsic value and perhaps of even greater importance than intrinsic value. Better, pragmatic holism may place the environment even higher than do intrinsic value theories. The holism of most of the pragmatists constitutes a challenge to value and ethics centered in the individual, and is more suitable to an environmental, as opposed to a humane, ethic. As the world is not devalued in the subject-object dichotomy for the pragmatists,[35] the possibility of valuing the nonhuman is less difficult and more consistent with its theories of value.

Applied to human experience of nature, pragmatic holism could be taken as the holism of the experience of nature. Nature would be taken as a holistic experience, as it is for some "deep ecologists," and not as an object.[36] Experience of nature taken as a whole is in a different class than the value of an object. Similarly, other living creatures could be viewed as sentient, not as "objects" as Descartes views them. If they were viewed as sentient, they would have more of a claim of moral standing or obligation. Further, moral standing could be detached from intrinsic value in a holistic theory, a detachment echoed by some environmentalists. Moral standing and obligation could be tied to environmental impact, that is, impact on the biosphere as a whole.

The environment as the context is an explicit part of the pragmatic world view.[37] The distinction of intrinsic value from instrumental value is in the world, and has consequences in the world. Pragmatism is defined in terms of consequences and is particularly cognizant of such actual consequences. Because impact on the environment is consequentialist in structure, pragmatism is most suited to provide a clear picture of such consequences and also a critique of harmful ones. The environment is the ultimate "context" of all of our lives, necessary for all the ends of life, whether growth, self-development, or satisfaction. Destruction of the conditions of all growth, human and nonhuman, is anti-pragmatic. The value of the environment for the well-being of humans and nonhumans is certainly a pragmatic consideration; this is not divorced from a scientific understanding of the environment.[38] Consequences can be known or predicted and evaluated; those with bad consequences could be evaluated as unjustified, their consequences without practical value to life or worse, harmful to life. Anticipating practical value in this sense implies that the world should somehow be improved as a consequence. A standard or norm for environmental impact could be derived utilizing this reasoning, a "melioristic" standard. Actions that destroy or degrade the environment are unjustified.

Superficially, experience ties pragmatism to the empiricism of Locke and thus to grounding in the subject. Although pragmatism was not conceived as subjective by James or Dewey, the use of the term "experience" as the framework for philosophy has unfortunate subjective connotations, which have led critics astray in their interpretations. Despite disclaimers, "experience" has Cartesian overtones, that is, raises the issue of who experiences? Unfortunately, "experience" may have subjective implications, which exclude the common world of animals, plants, and the environment. However, the use of experience as basic is not unanimous among the pragmatists. There are two alternatives to the metaphysics of experience. One is to eliminate the subject. The other is to bypass metaphysics. Both strategies have been used by lesser-known figures from the history of pragmatism. The first, Arthur Bentley, is tied directly to Dewey. The other figure, well known in his own day, has faded in fame. However, F. C. S. Schiller has provided an even more radical model, which may be relevant to environmental ethics. I will briefly discuss these more radical figures.

In his late years, Dewey collaborated with Arthur Bentley in *Knowing and the Known*. While the outlook in the latter work is consistent with Dewey's earlier philosophy, there is at least one major change from Dewey's earlier work. There is very little mention of "experience." Indeed, Dewey was almost apologetic to Bentley for having tried to rework the term experience by stipulating a new meaning.[39] Instead, the term 'experience' is superceded and "behavior" is substituted except in very exceptional circumstances.[40] In other words, Dewey himself saw that experience was problematic, since it had connotations that Dewey wished to avoid. I noted in the Preface to this book that *Knowing and the Known* is perhaps more Bentley's work than Dewey's. However, the larger point is that, whether primarily Bentley, or both Bentley and Dewey,[41] there is an alternative to the metaphysics of experience within the pragmatic tradition.[42]

"Transactionalism" may be described as a process metaphysic, that is, a metaphysic in which change or flux are considered basic, not 'substance.' Dewey and Bentley realized that the "subject" was simply a psychological form of the old substance and attribute metaphysic propounded by Aristotle. In the latter, substantialist view, a substance is hypostatized "behind" or "under" the event. For substantialists, the stream of consciousness in experience requires a "subject" who is, literally, 'thrown under' (sub-jectum) the stream as the substance who experiences.[43] However, the two pragmatists realized that this is at best a logical requirement: we never experience such a substance-subject. Similarly, "things" are conceived as substances by Aristotle and other substantialist metaphysicians. But this involves the hypostatization of a substance to "stand under" an event. Dewey and Bentley argued that this hypostatization is unnecessary, and could be eliminated, a la Ockham's razor. Transaction is viewed as replacing interaction between subject-substances. It is defined as "where systems of description and naming are employed to deal with aspects and phases of action, without final attribution to 'elements' or other presumptively detachable or independent 'entities,' 'essences,' 'realities,' and without isolation of presumptively detachable 'relations' from such detachable 'elements.'"[44] Transactionalism is radically anti-dualist.[45]

Dewey and Bentley explicitly reject any hypostatization of substances "under" events in *Knowing and the Known*. The flux or the process is taken as the first level of analysis, without any need to go behind events and look for—or invent—substances.[46] A "transaction" involves an event in an environment in which the event cannot be separated from the environment as a distinct "substance." 'Event' is also the term that connects *Knowing and the Known* with *Experience and Nature*. In the latter book, Dewey argued for a metaphysic of processes characterized by events. Similarly, *Knowing and the Known* outlines a metaphysic of process characterized by events, but without reference to experience. Rather, behavior is one process or event in the flux. Conceiving the organism as a distinct substance acting within an environment is viewed as "interaction." Transaction eliminates the reference to the organism as a substance. What is left is the event of the behavior of the organism in the environment. There is also a greater sense that the event of knowledge cannot be separated from an act of knowing.[47] Epistemological dualism is rejected: the knower is a part of the act of knowledge. This connects the views of Bentley and Dewey to those of F. C. S. Schiller, the English pragmatist whom I will cover next. As Winetrout has put it, "The knowing-known [is] taken as one process in which in older discussions, the knowing and the known are separated in interaction." For both Schiller and the transactionalists, the human element cannot be subtracted from the event of knowing. As Dewey remarks, the distinction of science and common sense is not based on the idea that "science is not a human concern, affair, occupation. For that is decidedly what it is."[48] It is not as if the act of knowing by humans is not itself an event in the flux.[49]

If anything, transactionalism is even better as a framework for environmental ethics than the superceded "interactionalism." Consider the definition of 'behavior,' a term that is to supersede 'experience,' in *Knowing and the Known*. "A behavior is always to be taken transactionally: i.e., never as of the organism alone, any more than of the environment alone, but always as of the organic-environmental situation, with organisms and environmental objects taken as equally its aspects."[50] In this view the organism is even more closely connected to an environment than in *Nature and Experience*. Behavior cannot be separated from an environment, interactionally. Both are involved in a transaction in which both are "aspects" of the same event or process. Since the metaphysics of the subject is the source of the devaluation of the environment, and the subject is treated as a detached substance, Dewey and Bentley's critique of substance, subject, and interaction provides a pragmatic alternative in which the "subject" is entirely superceded. Not even the residue of the subject in the form of experience remains outside of an environment. Humans cannot be separated from their environment and all of their transactions involve an environment as an inseparable aspect.

> Environment is not something around and about human activities in an external sense; it is their medium, or milieu, in the sense in which a medium is intermediate in the execution or carrying out of human activities as well as being the channel through which they move and the vehicle by which they go on. Narrowing the medium is the direct source of all unnecessary impoverishment in human living.

By cutting the ground out from under the subject, Dewey and Bentley have both undermined subjective theories of value and also enhanced the model in which value is an event within nature.

In this section I noted the movement in modern value theory away from natural value toward subjective value. The basis for the devaluation of nature was in a metaphysic of the subject for which intrinsic value was confined to the subject. Dewey was a critic of this approach and attempted to argue against subjective metaphysics, against humans as the great exception to nature. All the pragmatists argue for the interpenetrating of fact and value and thus the value of all objects of experience. This is one approach to overcoming the subjectivity of value in the modern tradition. However, there is a more radical possibility. It is to reject the necessary connection of value to metaphysics and thus subordination of value to metaphysics. This possibility has a pedigree going back to Plato's "Form of the Good." In this more radical step, value is conceived independently of metaphysics and even as its basis. Since environmental ethics is radical both in its conception of the scope of values and morals and its attempt to reformulate ethical principles, it constitutes a project for which the subordination of value to the subject is problematic. If value can be conceived as independent of and basic to the metaphysical subject, environmental ethics would have a more commanding theoretical standpoint. The shape of reality would be based on environmental values, not the reverse. Since one of the major theorists within environmental ethics, Tom Regan, has presented an axiological ethic, this more radical step could be supportive of his formulation of values independent of metaphysics.

Can value be conceived more radically as entirely independent of the subject? As basic to the metaphysics of the subject? Does pragmatism provide such a model? If value could be conceived as the ground or basis of the subject, then the subjectivization of value, which lies at the basis of the devaluation of nature, would be undermined. The subject would reflect the larger value relation that would be its essential ground. It was argued earlier that pragmatism was confused with more subjective philosophies because of the use of the term "experience" by many of its main figures, notably James, Lewis, and Dewey. However, a lesser-known figure from the history of pragmatism, the English pragmatist F. C. S. Schiller, used value itself as the basic framework of philosophy. This is indicated most clearly in his essay "The Ethical Basis of Metaphysics."[51] Ethics as the basis of metaphysics reverses the usual systematic relation among the main divisions of philosophy and places ethics in the position of a first philosophy.[52] For Schiller, the "good becomes a determinant of both the true and the real."[53] Because the world is incomplete, it can be shaped and this shaping can be regulated by ends, values, and moral norms. Values are the ground of how the world is shaped; ethics is neither a metaphysic nor based on metaphysics, but, in a radical reversal, at its basis. Thus a metaphysic of the subject would be derivative from ethics, not basic to ethics. Schiller describes his philosophy as "an assertion of the sway of human valuations over every region of our experience, and a denial that such valuation can validly be eliminated from the contemplation of any reality we know."[54] Indeed, Schiller argued for this position in the historical context of British idealism, which was a metaphysic of the subject. In turn ethics is concerned with the human good.

This is clearly an anthropocentric view, a "humanistic pragmatism." However, by extension, it could be transformed into the good of life, which would include the nonhuman in the scope of value and thus make life the basis of metaphysics. Values and morals together are ethics in Schiller's view, but ethics considers "the nature of life as a whole."[55] This view of ethics is close to Dewey's view and connects life as the bearer of value, nature, and holism. Since Schiller was also a value pluralist, who located value in different kinds, the value of nonhuman life could be fit into his philosophy. Further, the environment as a whole could be considered essential to human life and thus the ethical basis for shaping reality, including subjective reality. The good of the human subject is dependent on the good of the health of the environment. No higher human values are feasible without moral consideration of impact on the environment.[56]

Schiller's view of values, in which values form a replacement for metaphysical first philosophy, is still another alternative within the pragmatic tradition in which the dominant assumptions in the literature of environmental ethics are challenged. Dewey challenges the idea that ethics can only be based on intrinsic value, and also that the subject has autonomy from a natural organic environment. Schiller argues against the idea that intrinsic value is the quality of an entity and thus against the *pros hen* relation in Aristotle's treatment of value. Value does not require a basis in ontological metaphysics, and indeed can be viewed on the contrary as the basis of what "is." The inversion is in the relation of practice to theory, in that Aristotle took theory as primary, as first. The reduction of ethics to metaphysics in the tradition makes ethics subordinate to a foundational theory of knowledge. The pragmatists argue for the primacy of practice and thus of value. The practical world reflects value, not theory, and the objects of theory are first made by practice.

The primary role for ethics in Schiller's philosophy could provide a potent model for environmental ethics. The scope of environmental ethics gives everything at least potential moral considerability. Expansion of the moral sphere beyond the human to the nonhuman is based on the 'ought' and other ideals that regulate human action in line with the principle of their impact on the environment. Schiller provides such a model: ethics is not confined to the metaphysical or the ontological as the ought goes beyond the is as its basis. This is most suitable to environmental ethics as the latter expands the field of moral consideration beyond what is to what should be.

Notes

I will cite the following works of Dewey, which will be referred to by their initial date of publication. These will be followed by the University of Illinois edition of the Complete Works in brackets followed by the volume and page number of the latter where appropriate. For example, "Middle Works" Vol. 4 page 18 would be [M4: 18]. See *John Dewey, The Middle Works* and *John Dewey, The Later Works*, ed. by Boydston, J. (Carbondale: So. Illinois Univ. Press, 1984).

Dewey, John, *Ethics* (New York: Henry Holt, 1908, 1932 revision) co-authored with J. Tufts [L7; the 1908 ed. is M5].

———, *Reconstruction in Philosophy* (Boston: Beacon, 1920, 1972) [M12: 77–204].

———, *Human Nature and Conduct* (New York: Modern Library, 1922, 1957) [M14].

———, *Experience and Nature* (New York: Dover, 1925 [1958 ed., which follows the 1929 "second edition"]) [L1].

———, *The Public and Its Problems* (Chicago: Swallow Press, 1927, 1953) [L2].

———, *The Quest for Certainty, A Study of the Relation of Knowledge and Action* (New York: Capricorn/ Putnam, 1929, 1960 ed.) [L4].

———, *Freedom and Culture* (New York: Capricorn Books, 1939) [L13].

———, "Theory of Valuation," from the *International Encyclopedia of Unified Science* (Chicago: Univ. of Chicago Press, Vol. II, #4, 1939) [L13: 189–254].

———, *Liberalism and Social Action*, [reprinted in *John Dewey, The Later Works*, ed. by Boydston, J. (Carbondale: So. Illinois Univ. Press, 1984), v. 11: 1–68].

———, "Experience, Knowledge and Value, A Rejoinder," in Schilpp, P. A., ed. *The Philosophy of John Dewey* (New York: Tudor Publ., 1939, 1951), pp. 517–608.

———, "The Field of Value," in Lepley, R., ed., *Value, a Cooperative Inquiry* (New York: Columbia Univ. Press, 1949), ch. 3, p. 64 ff. [L16: 343–357].

John Dewey with Arthur Bentley, *Knowing and the Known*, reprinted in *John Dewey, The Later Works, Vol. 16, 1949–1952*, ed. by Boydston, J. (Carbondale: So. Illinois Univ. Press, 1989), pp. 1–280.

See also the essays "Value, Objective Reference and Criticism," *Phil. Rev.*, 34, 1925, 313–332; "The Ambiguity of Intrinsic Good," *Journ. Phil.*, 39, 1942, 328–330;

"Valuation Judgements and Immediate Quality," *Journ. Phil.*, 40, 1943, 309–317; and "Further as to Valuation as Judgment," *Journ. Phil.*, 40, 1943, 543–552. These are all reprinted as Part III of *The Problems of Men* (New York: Philosophical Library, 1946) along with the essay from Lepley and also in *John Dewey, The Later Works*, ed. by Boydston, J. (Carbondale: So. Illinois Univ. Press, 1984).

There are also a number of essays on value reprinted in vol. 15 of the Boydston edition of Dewey, *The Later Works*, including "The Meaning of Value," *Journal of Philosophy*, 22, Feb. 1925, [L15: pp. 126–133]; "The Ethics of Animal Experimentation," *Atlantic Monthly*, 138, Sept. 1926, pp. 343–346 [L15: 98–103]; and "Some Questions about Value," *Journal of Philosophy*, 41, August 1944, pp. 449–455 [L15: 101 ff.].

Preface

1. Bourke, V. J., *History of Ethics* (Garden City: Doubleday/Image, 1968, 1970). I am indebted to J. Kockelmans, *Contemporary European Ethics*, Introduction, for bringing this text to my attention.

2. These terms probably predated the rise of environmental ethics as a recognized field within philosophy. They have been used by several prominent figures within the field, e.g., Tom Regan and Bryan Norton. For Regan's and Norton's views see chapter one.

3. The articles calling for an environmental ethic will be covered in chapter one. The call for a new environmental ethic was probably first made by Richard Sylvan (Routley) in his well-known article, "Is There a Need for a New, an Environmental Ethic?" *Proceedings of the XV World Congress of Philosophy*, no. 1, Varna, Bulgaria, 1973, pp. 205–210. Reprinted in *Environmental Philosophy, from Animal Rights to Radical Ecology*, ed. Zimmernan, Callicott, Sessions, Warren and Clark (Upper Saddle River, NJ: Prentice-Hall, 1993/1998).

4. These articles will be covered in detail following chapter one.

5. Lumping all pragmatists together is a serious distortion of the pragmatic movement, which has resulted in a misreading of pragmatism with respect to the environment. Not only is there an internal debate within the pragmatic movement on issues of value, but the more idealist wing upholds a strong notion of intrinsic value. I will briefly discuss the pragmatists other than Dewey in the Epilogue.

6. Apparently, Richard Routley began the extension of the term "intrinsic value" to the nonhuman sphere as an issue or problem for environmental ethics. My source on this point is J. Baird Callicott, *In Defense of the Land Ethic, Essays in Environmental Philosophy* (Albany: SUNY Press, 1989), ch. 9, p. 157.

7. Callicott and Rolston. See chapter one.

8. The pragmatists differ the most on the last point, although each of their accounts also differs from the Utilitarian formulation of the relation in which duty is defined in terms of value. For Dewey, ethics is consequentialist, but consequences are one, albeit important, factor in a whole. For Lewis, ethics are the condition of value. For Schiller, ethics are bound up with consequences. These points will be examined in more detail in a projected future volume, *Pragmatic Consequentialism*.

9. The division of Dewey's voluminous writings by Jo Ann Boydston into the

early, middle, and late periods in the well-known complete edition of his works put out by Southern Illinois University Press does not connote periods in his intellectual life. It is more a matter of editorial arrangement by rough dates, I believe.

10. I am indebted to an anonymous reader at SUNY Press for raising this issue. Another question is whether other works from Dewey's very late years should be read as transactional, e.g., the important essay "The Field of Value." (I am indebted to still another anonymous reader of chapter two, which has been published separately in *Environmental Ethics,* for pointing out the question in relation to this latter text). If so, does this affect any coherentist reading of Dewey's many writings?

11. This is the view of T. Z. Lavine, who wrote the introduction to Vol. 16 of *John Dewey; The Later Works,* 1949–1952. See the introduction to the vol., pp. ix ff, esp. p. xxvii ff. Lavine also agrees with those who argue that the work is primarily Bentley's (see n. 13). However, this raises as many issues as it settles, especially what did Dewey believe? Was he a transactionalist, or to what degree was he one?

12. Sidney Ratner, quoted in ibid., Lavine, p. xxvii.

13. J. F. Ward, *Language, Form and Inquiry: Arthur F. Bentley's Philosophy of Social Science* (Amherst: Univ. of Massachusetts Press, 1984), p. 206; quoted in ibid., Lavine. This view is echoed by Paul Kress who argued that Bentley "dominated" the exchange between the two men (ibid., p. xxvi). A more balanced view is that of McDermott, who argues that Dewey also got his own point of view across.

14. Only ch. 10 is Dewey's; 1, 8, and 9 are by Bentley alone while the others were written by Bentley and reworked by Dewey.

15. I am indebted to another anonymous reader at SUNY Press for addressing Rorty's reading of Dewey. However, I do not find Rorty's reading convincing, as I will argue.

16. See *Consequences of Pragmatism* (Minneapolis: University of Minnesota Press, 1982), p. 89, n. 23. Since transactionalism is also a rejection of the substance-attribute metaphysics of Aristotle, a point I will cover in the Epilogue, it constitutes a rejection of the metaphysics of being. Since Heidegger explicitly adopted this metaphysic in *Being and Time,* whose preface outlines the problematic of being *qua* being, an Aristotelian formulation, Rorty's attempt to link Dewey to Heidegger is even more far-fetched than the link to therapeutic philosophy.

17. Ibid., esp. ch. 4.

18. Epicurus, who also saw philosophy as therapeutic, argued for withdrawal from practical affairs in a quiet and solitary life of undisturbed tranquility. Wittgenstein argued that "philosophy changes nothing." Both these views would have been an anathema to the practical Dewey; indeed he criticized Epicureanism on just this score. Thus Dewey commends the Epicurean emphasis on present goods, but *not* its ethic of withdrawal (*Human Nature and Conduct,* IV, 1, p. 268 ff.).

19. Peirce called his philosophy "objective idealism" at one point whereas Dewey was a naturalist. Rorty is also naturalistic, like Dewey but is more like Epicurus in his rejection of mind (*Philosophy and the Mirror of Nature,* Princeton University Press, 1978, ch. 1) and his therapeutic notion of the value of philosophy (in ibid., and op. cit., *Consequences of Pragmatism, passim.*). Despite his attempt to eschew system, Rorty has expanded pragmatism on the materialist side. (I have presented my arguments that Rorty is actually systematic in *Radical Axiology: A First Philosophy of Values,* forthcoming in 2003 from Editions Rodopi/Value Inquiry Book Series, ch. 1.) Since it is unclear

what practical consequences, if any, we are to draw from Rorty's work, it is difficult to judge if it may be suitable as the basis for an environmental ethics. I will therefore not treat it in the Epilogue.

Chapter One

1. The number of authors who have defended some form of argument for the intrinsic value of nonhuman individuals, species, habitats, landscapes, ecosystems, or the biosphere as a whole is vast and growing. The three authors chosen are representative, but hardly exhaustive. Others include Schweitzer, A., *Civilization and Ethics*, trans. J. Naish (for a selection of the relevant arguments see *Animal Rights and Human Obligations*, ed. T. Regan and P. Singer, Englewood Cliffs: Prentice Hall, 1976, sect. 10); Rachels, J., "Darwin, Species and Morality," in *Animal Rights and Human Obligations*, part 3, sect. 3; Hartshorne, C., "The Rights of the Subhuman World," *Environmental Ethics* (hereafter *Env. Eth.*), I/1, 1979, p. 49; Godfrey-Smith, W., "The Value of Wilderness," *Env. Eth.*, I/4, 1979, p. 309; Gunn, A., "Why Should We Care about Rare Species," *Env. Eth.*, II/1, 1980, p. 17; Taylor, Paul W., "The Ethics of Respect for Nature, *Env. Eth.*, III/3, 1981, p. 197, "In Defense of Biocentrism," ibid., V/3, 1983, p. 237. and "Are Humans Superior to Animals and Plants?" ibid., VI/2, 1984, p. 147; Miller, P., "Value as Richness: Toward a Value Theory for an Expanded Naturalism in Environmental Ethics," *Env. Eth.*, IV/2, 1982, p. 101.; Pluhar, "The Justification of an Environmental Ethic," *Env. Eth.*, V/1, 1983, p. 83; Hoff, "Kant's Invidious Humanism," ibid., p. 13; Lombardi, L., "Inherent Worth, Respect and Rights," ibid., V/3, 1983, p. 257.; Zimmerman, M., "Quantum Theory, Intrinsic Value and Pantheism," ibid., X/1, 1988, p. 3; O'Brien, J., "Teilhard's View of Nature and Some Implications for Environmental Ethics," ibid., X/4, 1988, p. 324; Merchant, C., "Environmental Ethics and Political Conflict: A View from California," ibid., XII/1, 1990, p. 45.; Anderson, J. C., "Moral Planes and Intrinsic Value," ibid., XIII/1, 1991, p. 49; Plumwood, V., "Ethics and Instrumentalism," ibid., XIII/2, 1991, p. 139.; Harlow, E., "The Human Face of Nature: Environmental Values and the Limits of Non-Anthropocentrism," ibid., XIV/1, 1992, p. 27; Booth, D., "The Economy and Ethics of Old Growth Forests," ibid., p. 43; Attfield, R., *The Ethics of Environmental Concern*, Columbia Univ. Press, 1983; and Katz, E., "Searching for Intrinsic Value," *Env. Eth.*, 1/3, 1987. For a history of the environmental movement that considers those who argue for the intrinsic value of the nonhuman outside of philosophy proper, e.g., in theology, see Nash, R. F., *The Rights of Nature* (Madison: University of Wisconsin Press, 1989).

2. See, e.g., Frondisi, R., *What Is Value—An Introduction to Axiology* (La Salle: Open Court Publ., 1971); Osborne, H., *Foundations of the Philosophy of Value* (Cambridge: Cambridge Univ. Press, 1933), Rescher, N., *Introduction to Value Theory* (Englewood Cliffs: Prentice-Hall, 1969).

3. I am indebted on this point to Stephen Pepper. See "Observations on Value from an Analysis of Simple Appetition," in Lepley, R., ed., *Value, a Cooperative Inquiry* (New York: Columbia Univ. Press, 1949). Osborne, 1933, calls certain other theories relational as well.

4. *Webster's New College Dictionary*, 7th ed., 1972

5. Ibid.

6. I am indebted to J. Baird Callicott on this point, i.e., the term "locus" of value, which occurs in several of his essays. See, e.g., *In Defense of the Land Ethic* (Albany: SUNY Press, 1989), Introduction, p. 3. The question of locus for environmental ethics is: can only individuals be bearers of value? Can species, landscapes, and ecosystems be value bearers? This question is closely related to that of the bearer of value, but less ontological in a sense. A bearer could be an individual, but as representative of a species in which value is placed or located.

7. Locus and bearer are often used interchangeably, especially by nominalists, but the distinction is important to holists. "Bearer" of value is a term that goes back at least to Sheler and perhaps farther. For Scheler, see *Formalism in Ethics and a Non-formal Ethics of Values*, trans. Frings and Funk (Evanston: Northwestern Univ. Press, 1916, 1973).

8. The origin of this term is in the essay by Goodpaster, K., "On Being Morally Considerable," *Jour. Phil.* 75, 1978.

9. Regan, T., "The Case for Animal Rights," in op. cit., *Animal Rights and Human Obligations*, P. IV, sect 1, p. 105 ff. For Singers view see ibid., P III, sect. 1 and also *Animal Liberation* (New York Review Press, 1975).

10. *Env. Eth.*, III/2, 1981, pp. 19–34.

11. Ibid., p. 20.

12. Ibid., p. 19.

13. See Regan, T., "The Case for Animal Rights," in Regan and Singer, eds., *Animal Rights and Human Obligations*, p. 105 ff.

14. 1981, pp. 24–25.

15. Ibid., p. 21.

16. Ibid., p. 20.

17. This argument is made Frey, R. G., "Why Animals Lack Beliefs and Desires," in Regan and Singer, 1976 (repr. from "Rights, Interests Desires and Beliefs," *Amer. Phil. Quart.*, Vol. 16, July 1979, pp. 233–239).

18. 1981, p. 22. The sentience argument is advanced by, inter alia, Singer in loc. cit.

19. Ibid., Regan, p. 24.

20. Regan prefers the term "inherent" to intrinsic, although he does not state why. I will presume that he means the same thing, as contextually there is no basis for a distinction.

21. 1981, p. 30.

22. 1976b, ("The Case for Animal Rights").

23. 1981, p. 31.

24. Moore, G. E., *Principia Ethica* (Cambridge: Cambridge Univ. Press, 1903), ch. 1.

25. 1981, Regan.

26. 1976b, p. 111. I will return to this point.

27. "Animal Rights and Human Wrongs," *Env. Eth.*, II/2, 1980, p. 99 ff. Cf. 1981, p. 30.

28. This includes most of the philosophers in the modern tradition, both rationalists (Kant's rational subject) and empiricists (Hume's moral sentiments and Mill's human happiness).

29. 1980, p. 115.

30. Ibid., p. 116.

31. 1981, p. 31.

32. See "Why Death Does Harm to Animals," in Regan and Singer, eds., 1976c, p. 153 ff.

33. See 1976b, pp. 110–111, and 1981, p. 31.

34. Callicott, J.Baird, "Animal Liberation a Triangular Affair," *Env. Eth.*, 2/4, 1980, p. 316, n. 17. (Reprinted in Callicott, 1989, ch. 1.)

35. These arguments are presented especially in Regan, 1976b.

36. Cf. the "ecofeminist" appeal to feelings as ethical warrants in Marti Kheel's "Nature and Feminist Sensitivity," p. 256 ff. in *Animal Rights and Human Obligations* (Englewood Cliffs: Prentice Hall, 1976). Kheel ignores nonfeminist appeals to feeling in ethics, e.g., the emotivist ethics of Ayer and Stevenson, the moral sentiments theory of Hume, as well as the role emotions play in intuiting value in Scheler's ethics.

37. For the notion of an axiological ethics see Findlay, J. N., *Axiological Ethics* (London: Macmillan, 1970). Also, cf. Kockelmans, ed., *Contemporary European Ethics* (Garden City, NY: Doubleday/Anchor, 1972), P. I.

38. See *Principles of Morals and Legislation*, ch. 17, sect 1, quoted in Regan and Singer, eds., 1976, p. 25. The extension of value and rights beyond the human sphere is treated historically in Nash, R. F., *The Rights of Nature* (Madison: University of Wisconsin Press, 1989).

39. *Lectures on Ethics*, p. 239, quoted in ibid., Regan and Singer, p. 24.

40. See *Groundwork of the Metaphysics of Morals*.

41. Aesthetic theories of intrinsic value have been defended by Plato, Pierce, Prall, and others. Other esthetic theories are included in the volume edited by Lepley, *Value, A Cooperative Inquiry* (New York: Columbia University Press, 1949).

42. This point requires some qualification. In a sense, as I will document, Callicott does argue from a subject.

43. 1976b, p. 111.

44. 1989.

45. 1981, p. 31

46. 1976b, p. 112.

47. Ibid., p. 113. The various particular obligations drawn by environmental ethicists are beyond the scope of this book. They will be covered only as germane to our topic.

48. 1989. Regan has shot back that Callicott is an "ecofascist" for not recognizing individual rights.

49. Of course there are implicit criticisms of the whole enterprise of assigning inherent value to the nonhuman sphere. These will be covered later in the chapter.

50. In Callicott, 1989.

51. Callicott, *In Defense of the Land Ethic, Essays in Environmental Philosophy* (Albany: SUNY Press, 1989), Introduction, p. 2. Actually, the environment is treated as instrumentally valuable by most theories, not as value neutral.

52. The qualification "moral" is important, as obviously Western philosophy as a whole cannot be accused of anthropocentricity. Lucretius's *De Rerum Natura*, for example, deals very little with humans. It is specifically moral philosophy that is anthropocentric.

53. Callicott, "Non Anthropocentric Value Theory and Environmental Ethics," *Amer. Phil Quart.*, 21/4, 1984, p. 299. Cf. "On the Intrinsic Value of Nonhuman Species," ch. 8 of 1989, p. 131.

Callicott relates in this regard a case made by Routley against a purely anthropocentric ethics. If atomic power were to render all human beings sterile and the humans in anger and frustration set about killing off all life on earth, there would be nothing morally wrong with such actions for an anthropocentric ethics (The "Last Man" Argument).

54. In "Intrinsic Value, Quantum Theory and Environmental Ethics," ch. 9 of 1989, p. 157 ff., second sect.

55. *Discourse on Method*.

56. 1989, ch. 9, p. 172.

57. Callicott notes Regan's classification of a prudential ethic as a management ethic in 1984, n. 3.

58. 1989, Introduction.

59. See esp. "Traditional American Indian and Western European Attitudes Toward Nature: An Overview," in 1989, ch. 10.

60. 1984, p. 299

61. Callicott's position has changed over the course of his career. His self-acknowledged "stridency" and strict application of the land ethic has mellowed such that he now seeks a common front with opponents within the environmental movement who are less radical. See "The Case Against Moral Pluralism," *Env. Eth*. 12/2, 1990, for the announcement of this change.

62. 1989, Introduction.

63. "The Metaphysical Implications of Ecology," in 1989, ch. 6, p. 101 ff.

64. Loc. cit., ch. 9 of 1989, p. 172.

65. Ibid., Introduction, p. 11. The sense in which evolution is used as a foundation will be covered later.

66. Ibid., ch. 4, "Elements of an Environmental Ethic: Moral Considerability and the Biotic Community," p. 63.

67. 1989, ch. 9. Callicott calls such an extension from self to nature "axiological complementarity" (p. 172).

68. Ibid., Introduction, p. 4. Callicott notes that the deep ecologists are even more radical as they criticize the whole relation of traditional philosophy with rationality.

69. Ibid., p. 5.

70. In one of his more speculative articles, he suggests that quantum mechanics might provide a basis for such an axiology, as the subject and object are not distinct in his reading of it. See ibid., ch. 9, p. 164 ff.

71. Ibid., ch. 8, "On the Intrinsic Value of Nonhuman Species," p. 146. Cf. ch. 2, "Review of Tom Regan, *The Case for Animal Rights*," p. 40; and 1989, p. 301

72. Ibid. (1984).

73. "Animal Liberation: a Triangular Affair," 1989 ch. 1 (repr. from *Env. Eth*. 2/4, 1980; hereafter 1980) p. 316. This view was modified in a later essay, ibid., ch. 4, "Elements of an Environmental Ethic: Moral Considerability and the Biotic Community," p. 63. Callicott, in a personal communication, has pointed out the evolution of his thinking. He thought that in my original draft I did not stress the historical devel-

opment of his views sufficiently. Therefore I modified this entire section to reflect where I believe he changed his views and which persisted. However, he may still believe that I have not done justice to them. While I am critical of some elements of his approach, I also greatly admire his pioneering efforts in environmental ethics, which has influenced my own views. The structure of this book is an attempt not to replace Callicott's views but provide them with what I believe to be a superior foundation in Dewey, rather than Hume. However, this effort could not have been undertaken without Callicott's work, which provided the framework.

74. Leopold's land ethic is considered a milestone by many environmentalists, not just Callicott, and has had an immense posthumous influence. It is presented in *A Sand County Almanac* (New York: Oxford University Press, 1949).

75. 1980, p. 319.

76. Ibid., p. 337.

77. 1984, p. 304.

78. Ibid. The qualification "present" is by contrast to Platonic-Leibnizian value theories, which Callicott believes could justify intrinsic value of the nonhuman, but not of concurrent species. See 1989d, "On the Intrinsic Value of Nonhuman Species," sect. 4 on "Holistic Rationalism."

79. Ibid., (1984). p. 305.

80. Ibid., p. 304.

81. 1989d, p. 130.

82. Ibid., p. 131.

83. Ibid. Cf. 1989, ch. 5, p. 98, where the definition is repeated almost verbatim.

84. E.g., Osborne, H, 1933, ch. 10, p. 88; Pepper, S., "Observations on Value from a Analysis of Simple Appetition," in Lepley, R., ed., *Value, A Cooperative Inquiry* (New York: Columbia University Press, 1949), p. 245. The relation of value requires a valuing subject in some relation to a valued object.

85. 1989, (Callicott), ch. 9, p. 158. The quote is a precis of Regan by Callicott, but in the latter's words.

86. Ibid., p. 161.

87. Ibid., ch. 10, p. 297, n. 8. This is a claim I will strongly contest in the Epilogue to this book. It is accurate to say that Lewis's theory regards inherent value as instrumental, but as Lewis clearly distinguishes intrinsic and inherent value, it is inaccurate to state that he has no intrinsic value theory. For Lewis's view see *An Analysis of Knowledge and Valuation*, B. III, ch. 12.

88. 1980, p. 325.

89. 1989, ch. 8, p. 136.

90. Ibid., ch. 9, p. 159.

91. Ibid., p. 158.

92. Ibid., ch. 8, p. 133. It is not clear how an extensionist, especially one as radical as Callicott, can coherently maintain a notion such as "value-free." If even soil and air have value, what is left to be value free?

93. Ibid.

94. Ibid., ch. 9, p. 169.

95. Ibid., p. 170.

96. E.g., Regan's as outlined earlier. Utilitarians, e.g., Singer, have in the main defended this view.

97. E.g., Schweitzer's "reverence for life." Callicott lumps together voluntarist and interest theories in 1989, ch. 8. Because a will to live gives an "interest," voluntarism is classed as a species of the interest theory. See the sect. "Conativism." Callicott also argues that a reverence for life ethic is impossible, and even "anti-natural," as I noted earlier. For Schweitzer's view see op. cit., Regan and Singer, eds., *Animal Rights and Human Obligations*, P. I, ch. 10. Callicott considers Rolston a conativist in "The Case Against Moral Pluralism" (Callicott, 1990) although in the same essay he is lumped with the "Kantians."

98. Ibid., (Callicott, 1989), p. 131 and the sect on "Conativism."

99. Ibid., p. 151 and the sect on "Conativism." Cf. the definition of human as the 'rational animal.' Callicott also considers theocratically oriented views based on biblical "stewardship." While such views are not anthropocentric in a strict sense, they still provide a rationale for human "dominion" (ibid., ch. 1 and 8). For a counterargument, see Attfield, R., 1983, P. I.

100. Ibid., ch. 9, p. 158.

101. Ibid.

102. Ibid., in the section on "holistic rationalism;" and also in Callicott, 1980; and 1984, p. 302 ff. Callicott includes Leibniz as among those who hold this view as well as Peter Miller. This is, of course, Callicott's interpretation of these thinkers rather than my own. For references on Miller's view see n. 1.

103. A view that is defended as adequate as an environmental ethic is that of the Native Americans. See Callicott, 1989, ch. 10–11.

104. Ibid. (ch. 8 of Callicott), p. 147.

105. Ibid., ch. 9, p. 160.

106. Ibid., p. 133. Cf. Levinas.

107. Ibid.

108. Ibid., ch. 9, p. 160.

109. See above and n. 86.

110. Ibid., ch. 8, pp. 133–134.

111. 1984, p. 305.

112. Ibid.; also in 1989, ch. 3; ch. 8, sect. on "Bio-empathy"; ch. 9, p. 160–163; and 1990. Callicott even ties in this naturalistic tradition to sociobiology in loc. cit., ch. 9. He believes the Ehrlich's rationale for intrinsic value may also be based on the moral sentiments view (ibid., 1989, ch. 8).

113. Callicott's summary of Darwin's position is taken from his reading of the *Descent of Man* in ibid. (1989) ch. 3, 8, and 9. It could be argued that Darwin conceives altruism as in the self-interest of the individual and thus that egoism is basic. Because this is a tangential point, I will not pursue it any further. Darwin argues in ibid. that "there is no fundamental difference between man and the higher mammals in their mental faculties." (Selection from ch. 3–4 in Regan and Singer, eds., 1976, Part I, sect. 10, p. 27).

114. Callicott, 1989, ch. 9, p. 163. Callicott notes that Leopold specifically cites Darwin in certain passages and was influenced by him in formulating the land ethic.

115. Ibid., ch. 3.

116. Ibid.

117. Ibid., ch. 8, p. 135. Cf. p. 161 on the value of species "for themselves" if not "in themselves."

118. Including Regan and J. Feinberg. The latter argues that only individuals can have interests, not species. For Callicott's consideration of Feinberg's view see ibid. For his arguments against Regan see ibid. and ch. 1 (1980).

119. Ibid., ch. 2, p. 42.

120. Ibid., ch. 6.

121. These points are covered in ibid., ch. 5.

122. 1980, p. 337.

123. Callicott discusses the value of landscapes in only one essay, "Aldo Leopold's Land Aesthetic" (ch. 13 of 1989). On the whole he rejects aesthetic intrinsic value as anthropocentric (ibid., ch. 8).

124. Ibid., ch. 7, p. 127. Cf. Pierce on "community."

125. 1980, p. 322.

126. This point echoes Aristotle's criticisms of Plato's theory of forms.

127. Callicott, 1990. Callicott argued for the separation of animal liberation and ecological ethics in 1989, ch. 1, but, in a later essay, called for their reconciliation (repr. in Callicott, 1989, ch. 3). However, the differences over what has intrinsic value and what is entitled to moral considerability remain.

128. Callicott, 1989, ch. 2, p. 40.

129. Callicott, 1984, p. 301; 1989, ch. 8, sect. "Conation."

130. Ibid. (1989).

131. Ibid.

132. Ibid., p. 147.

133. Although Callicott argues that granting wild animals human rights would be crypto-domestication, and "ludicrous" from the perspective of the Land Ethic, he has also called for reconciliation (see n. 127). He does argue that in the last analysis sacrifice of individuals to preserve the wild places as a whole is acceptable.

134. This expansion of ethical consciousness, first proposed by Bentham, was given an ideal reading by Lecky in *The History of European Morals* (quoted in Nash, 1989; for the history of this idea see Nash, 1989). Callicott argues that although Leopold's views have not been well received by many modern philosophers, due to the nominalistic bent of much of modern value theory, that he has included the idea of the expansion of moral considerability, the wider inclusion over time, in the land ethic (Callicott, 1989, ch. 5, p. 75).

135. Callicott, 1989, ch. 5.

136. Ibid., Introduction. For Leopold's views, see *A Sand County Almanac* (New York: Oxford Univ. Press, 1949), esp. pp. 201–203. For an essay that traces the development of Leopold's views see Norton, B., "The Constancy of Leopold's Land Ethic," in Katz and Light, 1996, ch. 5

137. Callicott, 1984, p. 305 ff.

138. Op. cit., Norton.

139. This parallels the debate over universals.

140. Callicott, 1989, ch. 8, p. 151.

141. Op. cit., Leopold, quoted in Callicott, 1989, ch. 8, p. 140.

142. Ibid., ch. 5 and Callicott 1990, pp. 121–123. Callicott argues, based on Darwin, that evolution is "slow and local," while human changes in the environment have been "abrupt and global." (Ibid., 1989, ch. 5).

143. Ibid.

144. Callicott 1990, loc. cit.

145. Ibid. This view stresses energy cycles, e.g., in photosynthesis and the food chain, rather than mechanical relations. Energy is more fundamental than matter in this view (Callicott, 1989, ch. 5).

146. Ibid., ch. 1.

147. The quote is from Leopold, op. cit., p. 202, quoted in ibid., ch. 5, p. 77.

148. Callicott's argument for this relation is that reason requires language, which in turn requires a "highly developed social matrix." Such a social matrix requires limitations on freedom in the form of an ethic. See ibid., p. 79 for Callicott's treatment of this point.

149. Ibid., Callicott, 1989, ch. 2; and ch. 8, sect. 2 p. 134 ff. In the latter section, he argues that rights are a modern notion, thus connected to Cartesian dualism. Appeals to rights don't occur in Asian and ancient writings.

150. Ibid., ch. 2, p. 42 ff. Callicott argues that predation selects out by an "invisible hand" the healthier individuals and thus is good for the species in the long run, thus tying Smith's laissez-faire economic views to Darwin. Against the utilitarian view that suffering is wrong, he argues that in the wild pain has a function, and that if "nature as a whole is good, then pain and death are good" (ibid., ch. 1) and that environmental health has "premium value" over comfort, pleasure, etc.

151. Ibid., ch. 8.

152. Ibid., p. 131.

153. Ibid., ch. 5, p. 80, emphasis in original. Callicott approvingly quotes Darwin: "All ethics so far evolved rests upon a single premise: that the individual is a member of a community of interdependent parts." (The quote is from the *Descent of Man*, pp. 202–203, quoted in ibid.)

154. Ibid., p. 77. Hare's view can be found in Hare, R. M., *The Language of Morals* (Oxford: Oxford Univ. Press, 1952) and *Freedom and Reason* (Oxford: Oxford Univ. Press, 1963).

155. Ibid. (Callicott) ch. 7.

156. For Moore's view see 1903, ch. 1.

157. Callicott argues that the normative element forms the minor premise in a practical syllogism, so that it is not derived from moral sentiments direct (ibid., Callicott, ch. 7). Moral sentiments are the major. However, this still is to derive an imperative for action from an 'is.'

158. Ibid., ch. 9.

159. Ibid., ch. 3, p. 58.

160. Ibid., ch. 4, p. 80. This conflicts with the predator-prey relations argument made by Callicott against individual intrinsic value, however. The norm that animals ought to be left alone is a better approach than the extinction argument; if we could genetically reconstruct rare species, the argument from eternal loss would not apply. But leaving animals alone might be requisite for other reasons, e.g., health of the habitat as a whole or the intrinsic value of the species.

161. Ibid., ch. 10, p. 164.

162. This phrase is derived from the author who proposed moral pluralism in environmental ethics, Christopher Stone. See Callicott, 1990.

163. Ibid.

164. In Callicott 1989, ch. 3.

165. Callicott, 1990, p. 113. Callicott rejects "deconstructionist" approaches to environmental ethics because of their rejection of metaphysical grounding.

166. Ibid., p. 123.

167. I am indebted to P. Taylor on this point. See Taylor, "The Ethics of Respect for Nature," *Env. Eth.*, 3/1, 1981. Callicott also fails to realize that his species ethic puts him in an Aristotelian metaphysic, not a Humean one.

168. Callicott, 1989, ch. 8.

169. I am not going to cover the problems of 'naturalism' vs. 'non-naturalism' with respect to Callicott as I agree with Osborne (1933) that the naturalistic "fallacy" is not a fallacy but a differing approach to values. I covered this issue to some extent in the section on the is-ought distinction; it seems to me that moral sentiments are a fact, and that Hume's is a naturalistic theory. This issue will be examined in more detail in ch. 2.

170. Rolston, Holmes III, *Environmental Ethics, Duties to and Values in the Natural World* (Philadelphia: Temple Univ. Press, 1988), Preface.

171. Ibid. On respect for nature cf. Taylor, P., 1981.

172. Ibid. (Rolston), ch. 1, p. 1. This view was anticipated by Dewey; see chap. 4.

173. Ibid., ch. 5, p. 160.

174. Ibid., ch. 1, sect. 1, p. 3; and ch. 9, p. 328. Rolston lists a number of such natural supports or conditions for culture, e.g., air, water, photosynthesis, etc. in loc. cit., ch. 1.

175. Ibid., ch. 1, p. 33.

176. Ibid., ch. 6, p. 192. This view was also anticipated by Dewey.

177. Almost. Rolston's use of "objective" to describe values actually involves the Cartesian standpoint of subject and object. However, he uses objective in contrast to subjective values.

178. Biocentric views are those that attribute value to all living things, e.g., Schweitzer's (see n. 1). Equalitarian views attribute value equally to all individuals, e.g., Regan and other animal liberationists as well as value nominalists.

179. Rolston, 1988, ch. 3, p. 116.

180. Ibid., ch. 1, p. 4.

181. Ibid., subsect. 2–3.

182. Ibid.

183. Ibid., p. 41 where he talks of our "conduct guided by nature." This is not neo-Stoical, as I will argue.

184. Ibid., ch. 9, p. 331.

185. Ibid.

186. Cf. Dewey's critique of the "spectator" view of mind and knowledge as the product of contemplation.

187. Ibid., ch. 1, p. 27.

188. Ibid., ch. 2, p. 82.

189. Ibid., ch. 4, p. 148.

190. Ibid., ch. 2, p. 58.

191. Ibid., ch. 1, subsect. 2.

192. See "Are Values in Nature Subjective or Objective," *Env. Eth.* 4/2, 1982, p. 134.

193. Ibid., pp. 146–147. Rolston also mentions "fitness to helps in the world," but this may mark instrumental value.
194. Rolston, 1988, p. 187.
195. Ibid.
196. Ibid., ch. 3, p. 100.
197. Ibid., p. 101.
198. Ibid.
199. This holds "even if . . . their behaviors work by genetic programs, biochemistries, instincts or stimulus-response mechanisms" (ibid., p. 109). The evidence is that they respond to stimuli, fight, run, grow and reproduce, and resist death.
200. Ibid., ch. 1, p. 32.
201. Ibid., ch. 6, p. 199. Cf. Nietzsche's view of nature (will-to-power) as destructive and constructive in the same moment. For Nietzsche's view see *Beyond Good and Evil*, sect. 1, #9.
202. Ibid., p. 15.
203. Ibid., ch. 6, p. 197.
204. Ibid., ch.1, p. 15.
205. Ibid., ch. 6, p. 198.
206. Ibid., ch. 5, p. 191.
207. Ibid., ch. 4, p. 150.
208. Ibid., ch. 2, p. 59.
209. Ibid., ch. 1, p. 28.
210. Ibid., ch. 3, p. 112.
211. Ibid., ch. 6, p. 231. The discovery of value seems to differ from that of fact or knowledge in that the process requires something more than internal representation, viz. internal "excitation." Value is not the conclusion of a logical argument or causal series, but is more direct (ibid., ch. 1, p. 28).
212. Ibid., p. 3.
213. Ibid., ch. 9, p. 335. Rolston argues that if intrinsic value "inheres in its holder, one will not need to assess the environment to justify intrinsic value" (ibid.)
214. This range of values is covered in ibid., ch. 7.
215. Qualities are examined in this sense by Aristotle (Categories) and Moore (*Principia Ethica*, ch. 1).
216. Ibid., ch. 5, p. 186.
217. Rolston, 1981.
218. Rolston, 1988, ch. 3, p. 100.
219. Ibid., ch. 1, sect. II, ch. 2, p. 52.
220. Ibid., ch. 3, p. 105.
221. Ibid., ch. 1, p. 3, ch.5, p. 187.
222. Ibid., ch. 3, p. 116. Cf. ch. 6, p. 197 where he notes dirt may not "have" intrinsic value.
223. Ibid., ch. 1 where human valuing adds new values to the natural whole and ch. 7, which lists such values by kinds.
224. Ibid., ch. 8, p. 302. If virtue *(arete)* is translated as excellence, which some more recent translations have used, then Rolston's naturalism approaches Aristotle's.
225. Ibid., ch. 2, p. 100.

226. This is connected by Rolston with Windelband's distinction of nomothetic and idiographic values, i.e., the value of the general and the singular in ibid., ch. 9, p. 344. For Windelband's view see his Rectorial Address, "History and Natural Science," trans. Oakes, in *History and Theory*, xix, 1980.

227. Ibid., ch. 1, p. 23. Rolston denies natural teleology, in the sense of the whole or design, but accepts it in individual organisms. The individual has goals and these may be embodied in the genetic set in the form of motivations (ch. 3, sect. 1). The genes are "evidently the property of the species," as they are passed down between and among individuals of the species (ibid., ch. 4, p.149). Callicott has pointed out that this echoes the theories of sociobiology. (For Callicott's view see "Rolston on Intrinsic Value: a Deconstruction," *Env. Eth.*, 14/2, 1992.)

228. Ibid., ch. 5, p. 187.

229. However, this view was anticipated by Dewey. See ch. 2. There is no evidence that Rolston has read Dewey in his texts.

230. Ibid., ch. 1, p. 4. Again, this view was anticipated by Dewey.

231. Ibid., ch. 7, p. 260.

232. Ibid., ch. 1.

233. Ibid., ch. 5, p. 105.

234. Ibid., ch. 1, p. 4.

235. Ibid.

236. Ibid., ch. 6, p. 224. Degrees of being or reality as corresponding to degrees of goodness or value can classically be found in the Platonic tradition, beginning with the *Republic*, VI, and also including Plotinus' *Enneads*, Augustine's *De Natura Boni*, and Boethius's *Consolation of Philosophy*. I am indebted to the late Prof. Reiner Schurmann on this point.

237. Callicott, 1988, ch. 6, pp. 207–208. This point was also anticipated by Dewey.

238. Ibid., ch. 6, pp. 222–224.

239. Ibid., ch. 1, p. 24. This may be referred to as "degree" of value.

240. Ibid., ch. 9. These are zones of differing kinds of "value generation," presumably artificial, horticultural, and natural.

241. Ibid., ch. 7.

242. Ibid., ch. 2, sect. 2. "We should not be blind to real differences between species, valuational differences that do count morally. A discriminating ethicist will insist on preserving the differing richness of valuational complexity." Rolston hints that humans are not of absolute value in his article "Feeding People versus Saving Nature," in Gottlieb, 1997, p. 208.

243. See ch. 8-9 of ibid. for Rolston's fairly thorough discussion of the ethics of priorities.

244. The arguments of Aristotle and Kant are in Regan and Singer (eds.), 1976, Part I, sect. 2 and 7.

245. Nietzsche argues that the lower is the condition of the higher, i.e., that a condition may be an instrument and lower in *Will to Power*, trans. Kaufman (New York: Vintage, 1967).

246. Rolston, 1988, ch. 6, p. 214 ff.

247. Ibid., p. 222.

248. Rolston considers the difficulty of defining species in the respect that is adequate to a non-individualistic value locus in several passages, e.g., ch. 4, p. 135 ff.

249. Ibid. Again, the similarity to Dewey is evident. Rolston notes, again like Dewey, the importance of growth to organisms, but argues that this is not some criterion of value or ultimate end. There are limits to healthy growth of organisms (ch. 8, p. 301).

250. Ibid., ch. 9, p. 336.

251. Ibid., ch. 6, p. 217.

252. Ibid., ch. 5, p. 187.

253. Ibid., ch. 6, p. 217.

254. Ibid., p. 220.

255. Ibid., p. 199.

256. This includes, as I have shown, the distinction of is and ought, fact and value, that has characterized the modern tradition, especially Hume and Kant, although there is an attenuation of the distinction rather than an abandonment of it in the notion that ought is discovered with is. For Rolston's discussion of this distinction, see ibid., Preface and ch. 3, p. 99. For his critique of value as confined to the conscious subject, see ibid., sect. 2 and op. cit., 1982, sect. 2–3, p. 127 ff. See also his denial that value should be limited to those with souls in ibid., ch. 2. For the role of (subjective) experience in valuation see ch. 1, sect. 2, p. 27 ff. For Rolston's critique of the locus of value in persons, of Kantian origin but echoed by Hampshire, Feinberg, and others, which Rolston views as a form of egoism, see 1988, ch. 4, p. 126 ff., and ch. 9, p. 340 ff. Rolston correctly denies that he is a Kantian, contra Callicott, 1990. For his views on relational theories, i.e., value as a relation of an object to a subject, including Lewis's theory, see 1982, p. 149.

257. For Rolston's critique of anthropocentrism see ibid., (1988), ch. 1, subsect. 2, p. 30 ff. and ch. 4, p. 154 ff. where he calls the anthropocentrism of humanistic ethics "morally naïve." For his critique of value as a human construct of "tertiary qualities," Alexander's theory, see ibid., ch. 6, p. 208, and 1982 sect. 2.

258. Nominalistic ethics, i.e., intrinsic value confined to individuals is criticized in general in ibid., ch. 4, p. 143 ff. Particular nominalistic theories, including interest, hedonistic and rights theories are criticized in, resp., ch. 1, p. 4 and 31; and ch. 2, p. 82.

259. Ibid., ch. 1, p. 41 ff.

260. Ibid., ch. 6, p. 213. For his critique of value as a human construct of "tertiary qualities," Alexander's theory, see n. 257.

261. Ibid., ch. 2, sect. 1

262. Ibid., ch. 5, p. 170. Rolston compares the natural community created in wild habitats to cultural communities, e.g. religious and scientific communities.

263. Ibid., ch. 1, sect. 3. Nietzsche also viewed nature as "indifferent," in *Beyond Good and Evil*, P I, sec. 9.

264. "Whatever has . . . resident value lays a claim on those who have standing as moral agents" in ibid., (Rolston), ch. 3, p. 96.

265. This connection is made by idealist-deontological theories, such as Kant's, perfectionist-virtue theories such as Aristotle's, and utilitarian-consequentialist theories, i.e., all the main traditions of Western ethics, in one form or another. Dewey's ethics are an exception, a point that will be covered in ch. 3 and 4.

266. An exception may be constituted by theories that posit intrinsic good in each human but a higher good in the divine.

267. Ibid., ch. 2, p. 59 ff. "Nature is not a moral agent." The pain caused to animals is acceptable in this view because it duplicates wild processes, so long as it is not unduly tortuous or based on gratuitous human values.

268. Ibid., ch. 5, p. 168 ff.
269. Ibid., ch. 2, pp. 52–57. Rolston accepts predation and limits the rights of animals to not suffering, thus coinciding with the utilitarian view.
270. Ibid., ch. 9, p. 334. Again Rolston's view does not resemble Kant's. Cf. Callicott 1990.
271. Ibid., ch. 3, p. 121 ff.
272. Ibid., ch. 4, p. 154.
273. Ibid., ch. 6, p. 224.
274. Ibid., p. 225.
275. Ibid., ch. 4, p. 140 ff.
276. Ibid., and also Gottlieb, 1997, ch. 11.
277. Ibid. (1988), ch. 4. Anthropocentric theories of obligation are adjudged self-serving and exploitative.
278. Taylor, 1981. Callicott's grouping of Taylor and Rolston is incorrect in this respect (see n. 256).
279. Rolston, 1988, ch. 3, p. 120.
280. Ibid., ch. 2.
281. Ibid., ch. 6, p. 228 and ch. 8, p. 271.
282. Ibid., ch. 7, p. 260 ff.
283. Ibid., ch. 4.
284. Ibid., ch. 5.
285. Ibid., ch. 8.
286. Ibid., ch. 7, p. 246 ff.
287. Op. cit., Callicott, 1992.
288. Cf. notes 256, 270 and 278.
289. See p. 33.
290. 1988, ch. 5, p. 191.
291. Ibid., ch. 6, p. 231.
292. It is secondary as value is the basis of obligation, but is in turn based on nature.
293. Ibid., ch. 3, p. 114.
294. Norton has acknowledged his debt to Peirce and Dewey in his article, "Integration or Reduction, Two Approaches to Environmental Values," repr. in Katz, E., and Light, A. eds., *Environmental Pragmatism* (New York: Routledge, 1996), section 6. In the sense that he sees theory as derivative from practice he is indeed pragmatic. However, his defense of subjective value is contrary to Dewey and even to Peirce, who called his philosophy "objective idealism." It is closer, I think, to the humanistic pragmatism of F. C. S. Schiller in one sense, since value is treated as anthropocentric (Schiller argued for "humanistic" pragmatism). However, Schiller also argued against grounding of value in the subject, or indeed grounding of value, a point I will cover in the Epilogue. Subjectivity of value presumes a Cartesian universe, something that all the classical pragmatists opposed without exception. Thus the question arises of whether "subjective value" is consistent with pragmatism as classically defined—i.e., whether Norton (and Weston, see next chapter) is truly a pragmatist. However, although this topic is beyond the scope of this book, I will address it briefly in the Epilogue.
295. Norton, B., *Why Preserve Natural Variety* (Princeton: Princeton Univ. Press, 1987), Preface. See also ch. 3, p. 46 ff. and ch. 11, p. 211 ff.

Notes to Chapter 1

296. Norton accepts the relational aspects of intrinsic value, which were outlined in the other three authors, viz., instrumental-intrinsic (ch. 1, ch. 8, p. 151) and means-ends (ch. 8, pp. 156–157). The reflexive aspect is hinted at in ch. 1, where intrinsic value is defined in terms of "the worth objects have in their own right."

297. Ibid., ch. 8, p. 158. Norton considers but rejects the distinction between is and ought and notes Dewey's arguments against value neutral knowledge in ibid., ch. 1.

298. Ibid., ch. 1, pp. 3–6 and ch. 8, pp. 158–159. He even agrees with what the ontology of value would be like if it were valid, i.e., value "of" an object, or "attributing" value "characteristics" to an object. See ibid., p. 152 and 159 n.

299. Ibid., p. 6.

300. Ibid., ch. 7, p. 135.

301. Ibid., ch. 8, p. 156 ff.

302. Ibid., ch. 1, n. 7.

303. Ibid., ch. 7, p. 135. Norton also questions whether impaired humans are moral beings, deserving of moral consideration on this premise, the "argument from marginal cases."

304. Ibid., ch. 10, p. 185.

305. Ibid., ch. 2, p. 43.

306. Ibid., ch. 5.

307. Ibid., ch. 1, p. 12.

308. Ibid., p. 10. On the instrumentality of demand and transformative values see ibid.

309. Ibid., ch. 10, p. 189.

310. Ibid.

311. Ibid., ch. 2, p. 25 ff. and ch. 11, p. 211 ff. Norton also argues the difficulty of assigning monetary price to "amenity values" in ch. 5.

312. Ibid., ch. 2, p. 25 ff.

313. Ibid., ch. 11.

314. Ibid., ch. 2, p. 43.

315. Ibid., ch. 10, p. 212. Norton himself argues against allowing any further extinctions in several passages, e.g., ch. 6, p. 122.

316. Ibid., ch. 11, p. 219.

317. Ibid., ch. 8, p. 152.

318. Ibid., p. 151.

319. Ibid., p. 156. Norton calls the first the "feasibility" condition, the second the "adequacy" condition. He notes as a qualification of the second that intrinsic value ascriptions may limit the *way* something is used but not its use as such (ch. 11, p. 219).

320. Ibid., ch. 8, p. 151 ff.

321. E.g., the discussion of rights in the context of interests in ibid.

322. Ibid., pp. 160–162.

323. See Taylor, 1981.

324. Unlike Rolston, Taylor extends the value of individuals to species and ecosystems based on the individual welfare of nonhuman animals. This is a reversal of Rolston's position on the relation of individual and more holistic values, because for Rolston ecosystemic value is prior to, productive of, and the condition of individual value. Further Norton claims Taylor leaves the character of intrinsic value "unspeci-

fied," a claim I would contest (see Taylor, 1981, p. 198 where the typical relational reflexive aspect is explicitly stated).

325. Norton, 1987, p. 162 ff.
326. Ibid., p. 159.
327. Ibid., p. 162 ff.
328. Ibid., ch. 9.
329. Ibid., p. 177 ff.
330. Ibid., ch. 2, p. 43.

331. In a sense the premise of extension of moral considerability makes species morality irrelevant. If animals, for example, have not been accorded moral standing in the past they should be now.

Prologue to Chapter Two

1. Taylor, Bob P., "John Dewey and Environmental Thought: A Response to Chaloupka," *Environmental Ethics*, XII/2, p. 175.

2. Katz, Eric, "Searching for Intrinsic Value: Pragmatism and Despair in Environmental Ethics," *Env. Eth.* IX/3, 1987, reprinted in *Environmental Pragmatism*, p. 307 ff. (Katz and Light eds., 1996).

3. Taylor, Bob P., "John Dewey and Environmental Thought: A Response to Chaloupka," *Environmental Ethics*, 12/2, 1990, pp. 175–184. Taylor notes that Chaloupka emphasizes Dewey's critique of dualism, stress on connections (thus holism) and diversity, and his naturalism.

4. Ibid., p. 180. For a contrary view see Rosenthal, S. and Buchholz, R., "How Pragmatism is an Environmental Ethic," in Katz and Light, 1996, ch. 2, which covers Dewey in general and Parker, K., "Pragmatism and Environmental Thought" (ibid., ch. 1), which covers Dewey's instrumentalist theory of value. Cf. also the later section on Weston below, which also invokes the instrumentality of values.

5. Ibid., p. 182. Taylor argues that Dewey's interest in science is also as an instrument for reshaping the environment for human purposes.

6. Ibid., p. 181.

7. Ibid., p. 177.

8. For example in comparing Dewey to Locke. While both are liberals, Dewey's twentieth-century liberalism is far removed from Locke's eighteenth-century form. Detailing the differences would be beyond the scope of this book, but Dewey's connection to Roosevelt's New Deal and his commendation of certain socialist proposals would hardly have been consistent with Locke's laissez-faire politics.

9. See C. A. Bowers, *Education, Cultural Myths, and the Ecological Crisis* (Albany: State University of New York Press, 1993), ch. 3, esp. pp. 88–104. I am indebted to an anonymous reader at SUNY Press for bringing this article to my attention.

10. Ibid., p. 97.

11. Ibid., p. 98.

12. Ibid., p. 100. Bowers seems to be a cultural relativist, arguing that any cultural tradition is a valid form of knowledge, at least if it allows a group to live in its environment. I wonder if he would propose to adopt cannibalism from New Guinea, since this might be a solution to overpopulation.

13. Ibid., pp. 103–104.

14. Dewey, J., *Freedom and Culture* (New York: Capricorn Books, 1939/1963 ed.), p. 27, emph. added. Cf. *Human Nature and Conduct* (New York: Henry Holt, 1922), Part One, sections 4–5: "Custom and Habit" and Custom and Morality."

15. Ibid., p. 29. Dewey's view of language is beyond the scope of this book. However, he was aware of the culture-language connection (ibid., ch. 2 passim). Consider his remarks at the end of chapter one of *Freedom and Culture:* "the state of culture is a state of interaction of many factors, the chief of which are [inter alia] . . . the arts of expression and communication." More, Dewey explored the nature-language connection at length in *Experience and Nature,* ch. 5.

16. *Ethics,* ch. 10, p. 171 of the 1942 ed. (New York: Henry Holt, 1932 revision). Dewey notes that the distinction of customary morality from reflective morality is not absolute, since the former may reflect past use of the latter.

17. This point has been stressed by several Dewey scholars, esp. Gouinlock. See Gouinlock, J., *John Dewey's Philosophy of Value* (New York.: Humanities Press, 1972) and also his "Dewey," ch. 10 of *Ethics in the History of Western Philosophy,* ed. Cavalier, R., Gouinlock, J. and Sterba, J. P. (New York: St. Martin's Press, 1989), pp. 306–332.

18. See *Freedom and Culture* ch. 4. Also, cf. ch. 6, "The Problem of Method," in *The Public and Its Problems* (New York: Henry Holt and Co., 1927, repr. by The Swallow Press, Chicago, 1954) in which Dewey argues against government by experts. This is an implicit critique of the role of a totalitarian party in ruling in the name of the masses, but without their consent, the program of Lenin in "What Is to be Done?" But it is also an implicit critique of the kind of technocratic society criticized by John McDermott in "Technology: The Opium of the Masses." (For the latter see *New York Review of Books,* July 31, 1969.)

19. Katz, E., "Searching for Intrinsic Value: Pragmatism and Despair in Environmental Ethics," in Katz and Light eds., 1996, ch. 15, p. 308 and 314. Reprinted from *Env. Eth.* 9/3, 1987. All citation is from the former, but the article will be referred to as Katz 1987, its original year of publication.

20. Ibid., p. 313.

21. Katz, "Imperialism and Environmentalism," in Gottlieb, ed., 1997, pp. 163–174, (ch. 8) 168. Katz grapples with the definition of nature in this article in the light of human nature, human defining of nature, and other problems of definition. Nature is not a "mere object" but a more active agent in his view.

22. Katz, 1987, p. 315.

23. Ibid. Katz notes that Weston, whom we will consider next, argues that pragmatism is subjective but not anthropocentric.

24. Ibid., p. 308.

25. Ibid., p. 314.

26. Ibid.

27. Katz, in Gottlieb, 1997, p. 168, and Katz, 1987, p. 311 ff.

28. Ibid., 1997.

29. Katz, 1987, p. 307.

30. Ibid., p. 311.

31. Ibid., p. 316 and in his "Reply" to Weston in *Env. Eth.*, 10/3, 1988 (repr. in Katz and Light, 1996.) He also argues that Weston, who will be covered later, overemphasizes the importance of the intrinsic value issue for environmental ethics (1987,

p. 308) and that Rolston attenuates the distinction between intrinsic and instrumental value, while Callicott has attempted to overcome the subject-object dichotomy at its root (ibid., p. 310).

32. This point will be documented in the last section of the book, the Epilogue.
33. Lepley, 1949, ch. 3.
34. The entire debate is reprinted from *Env. Eth.*, in Katz and Light, eds., 1996. Weston concentrates on the issue of intrinsic value in pragmatism, so has been chosen over other pragmatic environmental ethicists. For two of the latter, see n. 4.
35. Ibid., p. 285, and in "Before Environmental Ethics," ch. 7 of ibid., p. 139. (The latter is reprinted from *Env. Eth.*, 14/4, 1992. Citations will be from the former.)
36. Ibid., ch. 14, p. 285.
37. Ibid., p. 287 ff. For Moore's views see *Principia Ethica*, *Ethics*, et al. I would argue that Weston misinterprets Moore as a subjectivist, missing the holism of value in Moore of which experience is only a part. See *Princ. Eth.* Ch. 6.
38. I would contest all of these interpretations of Moore. In brief, the relation of means to ends constitutes intrinsic values, which for Moore are not grounded metaphysically (*Princ. Eth.*, Ch. 4) and are embodied in concrete instances. The theory may be abstract, but the bearer is not.
39. Ibid., (Katz and Light) p. 296.
40. Ibid., pp. 294–295.
41. Ibid., ch. 7, p. 141.
42. Ibid., ch. 14, p. 294.
43. Ibid., p. 293.
44. Ibid., p. 299, and ch. 7, pp. 139–140.
45. Ibid. (ch. 14).
46. Ch. 7 (1992), p. 139. This is in effect an argument for contingency and relativism of values, a conclusion Weston does not fail to draw. However, he argues that it is not relative within the evolving system of the West.
47. Ibid., ch. 14, p. 296.
48. Actually, the problem can be traced to Aristotle, with mention of it in Plato.

Chapter Two

For a list of Dewey's works, see the beginning of the Notes. As I noted, those will be referred to by their initial date of publication, followed by the University of Illinois edition of the Complete Works in brackets, followed by the page number of the latter where appropriate. For example, "Middle Works" Vol. 4 page 18 would be [M4: 18]. See *John Dewey, The Middle Works* and *John Dewey, The Later Works*, ed. by Boydston, J. (Carbondale: So. Illinois Univ. Press, 1984).

1. Katz, E., "Searching for Intrinsic Value: Pragmatism and Despair in Environmental Ethics," in Katz and Light, eds., *Environmental Pragmatism* (New York: Routledge 1996) ch. 15, p. 307 ff. Reprinted from *Environmental Ethics*, Vol. 9, #3, 1987; Taylor, Bob P., "John Dewey and Environmental Thought: A Response to Chaloupka," *Environmental Ethics*, Vol. 12, #2, p. 175. See the prologue for a more detailed statement of these views.

2. Dewey, *Experience and Nature* (Dover, 1925 (1958) p. 68 [L1: 61–62].
3. Ibid., ch. 1, p. 21 [L1: 28].
4. Ibid., ch. 2, p. 49 [L1: 49]. The notion of a "pluralistic universe" may be borrowed from James's well-known book with the same title. However, Dewey probably arrived at the same notion independently.
5. Ibid., ch. 7, p. 275 [L1: 210].
6. Ibid., p. 278 [L1: 212].
7. Nature includes both space and time, and nothing natural is outside of space or time.
8. *Human Nature and Conduct* (New York: Modern Library, 1922, (1957), hereafter 1922 [M14]), I, 3, p. 53.
9. Dewey argues that while philosophers have often been concerned with what is, natural science recognizes "processes" as universal, not 'being' (*Reconstruction in Philosophy*, Boston: Beacon, 1920, 1972, hereafter 1920, Introduction, p. xiii [M12]). Being is substance for Aristotle; form for Plato. Both of these emphasize the fixed over the changing in Dewey's view.
10. 1925, Preface, p. xi, ch. 2, p. 62 and 70 [L1: pp. ix, 57 and 63].
11. Ibid., ch. 3, p. 111 [L1: 92–93]. Nature viewed as mechanical, the Cartesian view, is a bizarre form of the cause effect relation taken as primary. It is bizarre as it takes a human artifice as the model for nature. Nature as artifice is precisely the opposite of both the ancient and the commonsense views of nature, which contrast the artificial and the natural. Dewey rejects the identification of nature with mechanism (ibid., ch. 2, p. 58 [L1: 54] and ch. 3, p. 98 [54]) which is tied in his view to dualism.
12. Dewey, "Theory of Valuation," from the *International Encyclopedia of Unified Science* (Chicago: Univ. of Chicago Press, Vol. II, #4, 1939) hereafter 1939, sect. 6, p. 423 [L13].
13. 1925, ch. 1, p. 59 [L1: 55]. Cf. ch. 3, p. 99 [84], where causality is characterized as this sequential order and events are seen as belonging in a history. However, this should not be confused with progressive theories of history. Natural history, contrary to Bowers, is not read as progressive, but as a history of events. Progress implies an end in the form of improvement, but *natural* history for Dewey is a stream of events without an end. For Bowers' view see *Education, Cultural Myths, and the Ecological Crisis* (Albany: SUNY Press, 1993) ch. 2, p. 95 ff.
14. Ibid., Preface, p. xii. For nature as a process see ibid., ch. 3, pp. 102 and 110 [L1: 86, 91].
15. Dewey argues against natural teleology as opposed to natural finality in *The Quest for Certainty, A Study of the Relation of Knowledge and Action* (New York: Capricorn/ Putnam, 1929) ch. 6, p. 100 [L4: 80–81]). He also argues against the identification of "natural ends with good and perfection" (1925, ch. 10, p. 395 [L1: 295]). Both these positions are implicit critiques of the Aristotelian identification of ends, good and natural changes.
16. 1925, ch. 2 [L1].
17. Ibid., p. 45 [45].
18. Ibid., ch. 3, p. 109 [91].
19. Ibid., ch. 2, p. 60 [56]. Such principles are generic characteristics of experience of events, i.e. part of the metaphysics of nature.

20. Ibid., ch. 3, pp. 97–98 [83]. These qualities are regarded as essential to existence: "those irreducible, infinitely plural, undefinable and indescribable qualities a thing must have in order to be" (ibid., p. 85 [74]).

21. Dewey contrasts the stable with the contingent in ibid., ch. 2, p. 60 [56]. Other words for the precarious are the "uncertain" and the "hazardous."

22. For nature as potentialities see ibid., ch. 7, p. 262 [201] and 1929, ch. 4, p. 100 [L4: 81]. Potentialities are "starting points" not "final ends."

23. Ibid. (1925), Preface, p. xv.

24. Ibid., ch. 7, p. 277 [L1: 211]. Cf. ch. 10, p. 424 [317].

25. Thus for Dewey scientific refinements are not primary, but based on ordinary experience. Cf. n. 2.

26. 1925, Preface, p. xi [L1]. Thus Dewey can treat art as the culminating event within nature and not as an artificial product.

27. Impulse and instinct are discussed at length in 1922, section II [M14]. Dewey argues the novel thesis that for humans, instinct is not original, as babies are trained into acculturation from birth.

28. 1922, II, 3, p. 105 [M14]. Thus contrary to Bowers (loc. cit., n. 13), nature is almost never unmediated by the customs and habits created by culture. What Dewey recognizes, however, is that different cultures are themselves responses to different natural environments.

29. Ibid., II, 4, p. 118

30. Further, the environment changes in the process of organic activity upon it, thus there are no "natural" impulses. Fear, e.g., must be plastic to be of any value for changing circumstances. Fear of trains cannot be inborn in the genetic sense if human genes predated trains.

31. 1925, ch. 6, p. 242 [L1: 186].

32. As noted earlier (n. 27) impulse and instinct are discussed at length in 1922, section II [M14]. Dewey basically rejects the idea of rationality as the specific difference of humans, arguing like Hobbes that reason is something that is achieved, not native, and that it is not always operative in humans. What is common to human nature is a raw set of potentialities, referred to as "native human nature." These native potentialities include impulse and certain instincts, which humans as "living creatures" possess. "Man is a creature of habit, not of reason nor . . . instinct."

33. Ibid., Preface, p. xiv. Dewey argues against the notion that life could be caused by matter, the materialist view, because both matter and life are seen as among the "events" of nature. "Not matter but the natural events having matter as a character, 'cause' life and mind" (ibid., ch. 7, p. 264 [203]). Life and mind are complementary characters of "the highly complex and extensive interaction of events" that is, they arose by historical processes within nature. Life is consistent with nature viewed as events.

34. 1922, IV, 4, p. 294 [M14].

35. 1925a, ch. 2, p. 67 [L1: 61].

36. Ibid., ch. 7, p. 261 [200]. This point will be discussed in more detail later.

37. Ibid., ch. 7, p. 277 [211].

38. Ibid., ch. 4, p. 163 [130]. Cf. ch. 2, p. 70 [63] and ch. 7, p. 279 [213] where he argues against the separation of life from nature and mind from organic life.

39. Ibid., ch. 10, p. 421 [315].

40. See notes 1 and 13.

41. Cf. 1925, ch. 6, p. 225 [L1: 174], where the isolation of the transcendent ego from natural existence is seen as the source of a host of problems, such as how to get back in touch with the world again.

42. Ibid., ch. 2, p. 45 [46].

43. Ibid., p. 75 [67].

44. "The Field of Value," in Lepley, R., ed., *Value, a Cooperative Inquiry* (New York: Columbia Univ. Press, 1949, hereafter 1949 [L16]) sect. 3, p. 66 ff. [343 ff.]. Cf. p. 73. Also cf. 1929, ch. 4, p. 81 ff. [L4: 65 ff.] where scientific observation involves behavior and other activity, in the form of experimentation and this is judged a new theory of experience. Dewey was active during the behavioral revolution in psychology. As I will argue, he did not have a rigid or mechanistic view of behavior. He attempted to transcend subjectivity by linking experience to activity.

45. In *Radical Empiricism,* James argues that experience precedes the distinction of subjective and objective experience. Dewey makes a similar argument in 1925, ch. 1, p. 8 [L1: 18]. Thus while Rorty is correct that Dewey regretted using 'experience' as a basic concept for his philosophy, Dewey did not view experience in subjective terms. (For Rorty's view see the Introduction.)

46. Ibid., p. 4a [12].

47. Ibid.

48. Ibid., ch. 1, p. 2a [11].

49. Ibid., ch. 6, p. 232 [179].

50. This is similar to an Aristotelian argument in the *Metaphysics* that sensations cannot be sensations of themselves, but indicate something is being sensed. Dewey argues that ultimately dualism ends in solipsism as the connection of mind and matter cannot be made if they are distinct substances. Experience is shut off from nature conceived as purely mechanical.

51. Dewey's naturalistic method is an "empirical method" based on primary experience, which remains true to nature. It is consistent with both science and nature as it eliminates anything inconsistent with the "nature of things" (1925, Preface, p. x).

52. Ibid., ch. 7, p. 272 [L1: 208]. This level is characterized by Dewey as "association, participation, communication" as well as "individualities." Mind is not autonomous or isolated, but is formed by a distinctive social and cultural environment. The formation of awareness is also self-consciousness, or individuality. However, this reflects a social and cultural milieu.

53. Ibid., p. 284 [217]. Dewey notes that there are different views of the relation of mind to nature and body in different cultures (ibid., p. 248 [191]).

54. Ibid., p. 285 [218].

55. 1922, Preface, p. ix [M14].

56. 1925, Preface, p. xiii [L1]. Dewey notes that feelings are also in nature. Feelings mark off animate from physical nature, but are still events within organic nature, which in turn is in the larger natural world (ibid., p. xiv). The kinds or aspects of experience are also in nature.

57. Ibid., ch. 2, p. 45 [46]. This quote may constitute Dewey's answer to the is-ought dichotomy, as, if things are as they should be, the is and the ought, fact, and value, can coincide. New goods and ideals are "brought about" through activity. Ideals brought about or realized are 'oughts' become 'is,' i.e., values made or facts. More, values cannot be outside of nature if nature is the whole. Cf. n. 109.

58. Dewey notes that modern thought has generally excluded value and purpose from the world of nature as a part of its general isolation of humans.

59. Ibid. (1925), p. 96 [L1: 82].

60. 1922, IV, 3, p. 286 [M14].

61. 1925, ch. 3, p. 104 [L1: 88].

62. For a discussion of the is-ought distinction in relation to the naturalistic fallacy see n. 109.

63. Ibid. (1925).

64. Ibid., p. 95 [81]. This is particularly true of the modern period, which relegates ends to consciousness. If ends are confined to consciousness, they are isolated from "objective nature."

65. Ibid., Preface, p. xi.

66. Ibid., p. 80 [70].

67. "That organization which is the 'final' value for each concrete situation of valuation thus forms part of the existential conditions . . . in further formation of desires and interests or valuations." (1939, sect. 6, p. 430 [L13]).

68. 1922, IV, 1, p. 262 [M14]. Fixed ends subordinate present change to a future goal, a common model for the tradition. However, ends are part of an ongoing process of change, not outside of that process. Evolution has meant continuity of change is a corollary part of this process, part of natural history. A merely stable world would have no consummations of experience, no peaks or valued, if precarious moments. Dewey believed that evolution had been misconstrued, e.g., by Spencer, who turned it into a doctrine of fixed ends.

69. 1925, ch. 2, p. 62 [L1: 57]. In this passage Dewey comes close to adopting Brogan's comparative theory of value. For a discussion of Brogan's theory see Mitchell, E. T., "Values, Valuing and Evaluation," in Lepley, 1949, p. 202 ff.

70. Dewey has often been read as denying intrinsic value, e.g., by Beardsley, M., "Intrinsic Value," *Philosophy and Phenomenological Research*, 26, 1965, p. 6 ff. However, this is a misreading, as Gouinlock, inter alia, has extensively documented. (For Gouinlock's view see *John Dewey's Philosophy of Value*, New York: Humanities Press, 1972). One need only read the first and last chapters of *Experience and Nature* to realize that Dewey did not reject intrinsic value, only foundational grounding in intrinsic value. I will argue this point in more detail in chapter four, "Dewey's Holism."

71. 1922, Preface, p. 11 [M14]. The relation of natural and social environments will be covered in a later section.

72. Ibid., III, 2, p. 176.

73. Thus impulses, like desires, must be plastic to cope with altered environments.

74. 1922, II, 5, p. 145 [M14].

75. Ibid., III, 4, p. 187.

76. Needs reflect a tension requiring an alteration of the relation to the environment (1925, ch. 7, p. 253 [L1: 195]).

77. 1922, III, 8, p. 231 [M14]. It is notable that habit must be adjusted to changes in the environment, despite the routine character of habit. Dewey notes that "adequate preparation" leads to "intelligent adaptation" (III, 9, p. 249).

78. 1925, ch. 1, p. 9 [L1: 19]. He also qualifies this inclusive model by noting that "higher organisms" act with reference to a "spread out environment." A large animal may need more space to roam than a plant.

79. Ibid., ch. 7, p. 283 [216].

80. 1922, I, 1, p. 23 [M14]. Knowledge is of instrumental value for this process.

81. Another factor is what Dewey refers to as the "social environment." This point is maintained throughout the corpus, from *Ethics* (1908, M5) to the late works (1925, ch. 6, p. 233 [L1: 180]). Again, this is not confined to humans; many other species are gregarious. Further, in the process of child rearing, parents of many species use association to give the young an advantage in the struggle for existence, comparable to human education. This social environment is part of the natural environment of many organisms, then, continuous with nature in a larger sense. The social environment for humans is the influence of the social life on a plastic human nature (1922 [M14]; 1925, Forward, p. 6 [L1: 17]). Morals, e.g., are inherited from the social group, often in the form of customs (1922, I, 4, p. 55 [M14]).

82. 1922, p. 60 [M14]. Again, Bowers' general criticism that Dewey ignores culture is fallacious. Culture is treated as the social environment. Bowers argues that Dewey preferred "social" to cultural. However, 'social' was actually generic for culture in Dewey. Each culture forms a specific social milieu; society is the general form of such cultures. (for Bowers' view see op. cit., n. 13, p. 96.)

83. 1925, ch. 6, p. 233 [L1: 180]. "Different customs shape the desires, beliefs, purposes" (of humans) (1922, I, 4, p. 60 [M14]).

84. 1925, ch. 5, p. 169 [L1: 134]. This anticipates Wittgenstein's similar view by several decades. Again, Bowers' contention that Dewey "failed to recognize the culture-language-thought connection" is contradicted by this quote.

85. Ibid., ch. 7, p. 258 [198].

86. Dewey notes that both are thus "means and attained result or 'end' in all cases of valuation" (1939, sect. 7, p. 433 [L13: 240]). Cf. 1925, ch. 7, p. 271 [L1: 207].

87. 1925 ch. 1, p. 3a [L1: 12]. Impulses are also shaped by the environment or "surroundings" (1922, II, 2, p. 91 [M14]).

88. 1925, ch. 7, p. 247 [L1: 190].

89. 1922, III, 1, p. 168 [M14].

90. The extension argument has been used by a number of environmental ethicists to attempt to include animals, plants, or all of nature in the scope of morals. Examples are T. Regan and P. Singer.

91. 1939, sect. 3, p. 396 [L13: 204]. Dewey rejects any "nonnatural" critique of desires and interests as desires and interests are "ineffectual" apart from environmental conditions (ibid., sect. 4, p. 408 [216]).

92. 1922, IV, 4, p. 248 [M14], emphasis in original.

93. Bowers also seems to confuse the social with the natural in his treatment of Dewey's philosophy of nature. Dewey argues that "until the integrity of morals with human nature and of both with the environment is recognized, we shall be deprived of the aid of past experience to cope with the most acute and deep problems of life" (ibid., Introduction, p. 13).

94. 1925, ch. 7, p. 281 [L1: 214–215].

95. By inference it is the individual organism, as experience is individual (1922, IV, 1, p. 269 [M14]) and experience is within the organism. However, this *prima facie* conclusion may be modified in the next section.

96. 1922, I, 4, p. 63 [M14]; cf. I, 2, p. 31. This formulation echoes a Kantian point from another context.

97. Ibid., I, 3, p. 42. Without habit the mind could be conceived as separated from the body and 'mental' different in kind (ibid., I, 2, p. 33).

98. Dewey argues that action does not require a preceding motive as humans "can't help acting" (1922, II, 3, p. 110 [M14]).

99. For "psycho-physical" see 1925, ch. 7, p. 255 [L1: 196]. For desire as "modes of action" see 1922, III, 8, p. 230 [M14], and 1939, sect. 3, p. 396 [L13: 204] and sect. 7, p. 433 [L13: 241]. For feeling (and reason) as springing up within action see 1922, I, 5, p. 71.

100. 1925, ch. 7, p. 293 [L1: 223].

101. Weston, A., "Beyond Intrinsic Value, Pragmatism in Environmental Ethics," in op. cit., *Environmental Pragmatism* (New York: Routledge, 1996) ch. 14, p. 285 (reprinted from *Env. Eth.*, 7/4, 1985). This point is also echoed by Bowers (see n. 13).

102. 1925, ch. 1, p. 16n [L1: 24]. Dewey is of course only one pragmatist, but the anti-Cartesian element is present in all the pragmatists in some degree, e.g., in Peirce's idea of the "community of inquirers." I will cover this point in more detail in the Epilogue. Katz's view may reflect the popular understanding of pragmatism as "whatever works," regardless of consequences, and thus as pro-technology at whatever cost, viz. to the environment. Clearly, this is not Dewey's view because he rejects the notion that ends justify any means (1922, [M14] passim) or the radical separation of ends and means.

103. 1939, sect. 8, p. 443 [L13: 251]. Cartesian dualism results in a "unnatural" self, which is alienated from the rest of nature (1925, ch. 1, p. 24 [L1: 31]).

104. Dewey is not even denying subjectivity in the sense of privacy or individuality. A personal self emerges out of social and organic interaction (1925, ch. 6. p. 208 [L1: 162). The subjective emerges out of primary experience, i.e., behavior. It is not a "given," as in the Cartesian analysis (ibid., ch. 1, p. 13 [22]). Desires and satisfactions are natural and "objective," not subjective and private (Ibid., ch. 2, p. 64 [59]).

105. For example, Dewey argues against perfection as a final goal, a foundation of all activity. "Not perfection as a final goal but the ever-enduring process of perfecting, maturing, refining is the aim in living" (1920, ch. 7, p. 177 [M12: 172 ff.). Satisfaction is the "recovery of equilibrium" as a consequence of changing the environment by the organism (1925, ch. 7, p. 253 [L1: 195]).

106. 1925, ch. 8, p. 303 [L1: 230]. Consciousness is awareness of meanings, esp. linguistic (ibid.).

107. 1922, III, 1, p. 166 [M14]. Thus experience is not the experience of a mind or consciousness or transcendental unity of apperception, etc. Cf. 1939, sect. 5, p. 429 [L13: 237] where he argues against a "mentalistic psychology" that separates mind from the body and the physical. Stimulus and response are also part of an ongoing process.

Habits are important as an instrument of value for Dewey since they embody past experience, and thus practical knowledge. This in turn is instrumental for coping with new situations with some means other than blind or momentary impulse. "The lost objective good persists in habit, but it can recur in objective form only through some condition of affairs which has not been yet experienced" (1922, I, 3, p. 53 [M14]). Morals are primarily "objective" as they are based on habit, which incorporates and adjusts the environment (ibid.). Dewey argues that a scientific ethics would switch attention from the subject to the object (1925, ch. 10, [L1]) and would involve the view of humans as agents, not ends. "Ends would be found in experienced enjoyment of the fruits of a transforming activity" (ibid.). A "false psychology" based on the isolation of a subject shuts morals off from habit in its "objective consequences" (1922, I, 3, p. 54 [M14]). This must be anticipated in the process of deliberation. Improvement in habits is intelligent as better adapted to the environment. Thus good is more than merely the attribute of an activity of an organism; it is an attribute of a process that the organism goes through in coping with the environment.

108. 1920, ch. 7, p. 162 [M12].

109. The ought or ideal can be "brought about" or realized in practice guided by value. This is continuous with the processes of nature, e.g., means to ends as causes to effects. It is not unnatural, although it may be secondarily natural. The new objects brought about using natural processes in inventive ways have value in nature. Thus natural good is not confined to antecedent nature, any more than the emergence of life out of the previous, merely physical environment is unnatural. Natural qualities are in nature for Dewey: value is an emergent quality of nature itself. Norms as social facts further attenuates the "is-ought" distinction. Dewey argues that morals are social not simply as an ought but as a fact: we are held accountable for obeying social strictures willy-nilly.

This constitutes Dewey's answer to the "naturalistic fallacy" since if ought and is can never coincide, if they have no connection, they must be in different universes. This is to speak nothing of the problems of maintaining the validity of the "naturalistic fallacy," e.g., that for Hume, value attaches to facts, viz., the facts of human subjective states such as passions. The argument is also based on the reality of "being" or ontology in the form of the "is" of the distinction, but value as an attribute of being is rejected by Dewey. Moore's contention that language must be followed in determining meaning is contradicted by his treatment of the good in terms of a "nonnatural" predicate in order to get around Hume's arguments. Good is attributed by language in value judgments that do not distinguish natural and nonnatural or good as nonnatural. Either language ought to be followed consistently or it is not a reliable guide in the first place, as Dewey argued.

110. 1949, p. 69 [L16: 348].

111. The term has been kidnapped from the 1973 translation of Scheler (1916) by Frings and Funk. See Scheler, M., *Formalism in Ethics and a Non-formal Ethics of Values*, trans. Frings and Funk (Evanston: Northwestern Univ. Press, 1916, 1973). For an example of grounding value in ontology see Hans Jonas, *The Imperative of Responsibility, In Search of an Ethics for the Technological Age* (Chicago: University of Chicago Press, 1984). For Dewey's characterization of value as primarily adjectival see 1949 [L16].

112. 1949, [L16: 343 ff.] passim. Dewey also argues that moral goods and ends exist "only when something has to be done" (1920, ch. 7, p. 119 [M12]).

113. 1949, p. 65 [L16: 344].

114. This may also represent different stages of Dewey's career and a shift in emphasis. Habit is the main term used in connection with activity in the earlier period, especially Dewey, 1922 [M14]; while "behavior" is emphasized in the later works, 1939 and 1949 [L13: 189 ff., and L16: 343 ff.]. However, both terms occur in both periods. Both bear specialized meanings for Dewey. Habits are the "demand for a certain kind

of activity," i.e., both a determinate sort and a fixed or settled way of handling such a kind. It suggests a "hold, command over us." Thus they are more forceful than attitudes or dispositions. Dewey uses the word habit in a wider sense than in common usage. He argues that this is needed to "express that kind of human activity which is influenced by prior activity and in that sense acquired." This sort of activity also "contains within itself a certain ordering or systematization of minor elements of action . . . is projective . . . ready for overt manifestation; and which is operative in some subdued subordinate form even when not obviously dominating activity" (1922, I, 2, p. 39 [M14]). Dewey also argues that there is no hard distinction between deliberate action and activity due to habit and impulse. Thus action seems to be different only in degree from habit and impulsive activity. However, habits can be routine and unreflective, i.e., not involving deliberate action or even consciousness. Habit and impulse are seen as the key to a non-introspective psychology by Dewey, a view that was soon to bear fruit in the behavioral revolution.

Habits are activities that reflect past experience, in Dewey's sense of behavioral experience, for they are acquired through it (ibid., I, 4, p. 62). Habits spring from the same "native instinct," but become diverse through interaction with different environments (ibid., II, 1, p. 87). These kinds of experiences have a hold on our present activity and can manifest themselves at any time in present actions. Thus they are compared to tools, as they can be used as means, but they are active means, with a hold over us (ibid., I, 2, p. 26). Habits are compared to will by Dewey in ruling over thoughts, forming desires, and furnishing means for activities (ibid.). They have a regulative role over conduct and the formation of goals and purposes. Habit and impulse are viewed as the primary determinants of conduct, a synonym of behavior (1922, III, 5, p. 206). "Our purposes and commands regarding action . . . come to us through the refracting medium of bodily and moral habits" (ibid., I, 2). Habits are formed in relation to ends in view through recurrent past activities, while impeded habit is one source of formation of the end in view. In other words, habits are teleological, not merely mechanical. They are required for knowledge of good as the true source of knowledge. "Only the man whose habits are already good can know what the good is" (ibid.). Adults know with their habits, according to Dewey, not with their consciousness (ibid., III, 2, p. 172). In sum, habits are regulated activities or conduct that reflects past experience. Since they were formed by reference to ends and purposes, they can be judged good or bad. More, they embody the experience of good and this has its affect on present activity. Good habits lend themselves to good activities; bad habits, to bad.

The relation of habit to other forms of activity, conduct, and behavior is that habits are regulative over present conduct and behavior, whether good or bad. Whereas behavior and conduct are mostly observable, however, habits may operate subconsciously, as past experiences that determine present conduct (ibid., I, 2, where habits are said to work "below direct consciousness." Cf. Freud. They represent the memory of behavior, so they were originally public and their manifestation in present behavior is also public. However, their influence over present behavior is not observable, except in the sense that the conduct is such a manifestation). Behavior and conduct are also defined teleologically by Dewey (ibid., III, 6, p. 207. Behavior is not defined as in physicalism but refers to "animal life processes.") Behavior is the field of value facts; the field of value consists in behavior involving "selection-rejection" (1949, p. 64 [L16: 343]). Valuations are patterns of behavior and the test for the existence of valuation is behavioral. "It is by observations of behavior . . . that the existence and description of valuations have to be determined" (1939, sect. 3, p. 395 [L13: 203]). Thus value is connected with certain kinds of behavior, those involving selection, not all behavior. Cf. Peirce on nature as "acquiring habits." Also cf. transactionalism, where behavior is accepted but not hypostatization of a substance behind the behavior. Organism is accepted however, and presumably the existence of habits exhibited in behavior.

115. 1925, ch. 8, p. 300 [L1: 228].

116. 1922, I, 6 [M14]. Dewey concedes the difficulty of developing a science based on or involving individual behavior (1939, sect. 4, p. 400 [L13: 208]). Nevertheless, desire and interest as constituents of value must be interpreted behaviorally (ibid., sect. 7, p. 433 [241]).

Although Dewey himself uses the term "objective" in this text, it is also subject to the qualification mentioned at the end of the last paragraph. Its use by Dewey is explained by the historical context in which he wrote, in which Cartesianism in one form or another was still the reigning paradigm.

117. 1922, I, 1, p. 17 [M14].

118. Ibid., III, 7, p. 222. Moral facts are tied to special courses of action (p. 224). Certain values are tied to individual lives, e.g., equality. "Equality of value has . . . to be measured in terms of the intrinsic life and growth of each individual" (1908/1932, p. 383 [L7: 345]).

119. 1939, sect. 8, p. 444 [L13: 252]. Dewey argues the point that subjective ethics is anti-social in 1922, IV, 1, p. 269 [M14]. "We can't help being individual selves" but this should not result in subordination of the common good to private. Thus Stevenson's charge that Dewey's ethic is "individualistic," a charge echoed by Bowers, is unjustified. Morality is social, and ethical warrant is not based on subjective individual states, individual values, or consequences for individuals.

120. Cf. 1920, ch. 7, p. 169 [M12: 172 ff.] where the "intellectual" value of drawing general classes of goods from comparable cases is mentioned to 1939, sect. 6, p. 424 [L13: 232], where general ideas of value are used as intellectual instrumentalities. They have no bearer except as an idea, so are not cases of fixed value. Nevertheless, they have instrumental value as qualities.

121. 1920, ch. 7, p. 166 [M12: 172 ff.].

122. 1939, sect. 1, p. 382–3 [L13: 190–191].

123. 1949, p. 67 [L16: 347].

124. That Dewey's instrumentalism includes overall ends, such as growth, self-development, and life can be shown by a careful reading of 1908/1932 [M5/L7], 1920 [M12] and 1922 [M14]. I will cover this point in more detail in ch. four.

125. 1949, p. 67 [L16: 347].

126. 1922, II, 5, p. 127 [M14]. Dewey argues that there is not a special "instinct" of self-preservation. Rather, are there specific responses to the stimulation of the environment. This constitutes an implicit challenge to the notion that animal behavior reflects "instincts," since specific responses, though they may have had their origin as instincts, are plastic and adaptable enough to deal with novel situations. This includes animal responses; animals also encounter new situations that genetic programming could not have anticipated. Behavior to preserve life does not imply an "instinct" for self-preservation.

127. 1925, ch. 7, p. 281 [L1: 214].

128. Ibid., p. 282 [215]. Dewey does not think that this is a good argument for thought in "lower organisms," however.

129. Ibid., ch. 1, p. 23 [30].

130. Ibid., ch. 7, p, 256 [197].

131. Ibid., Preface, p. xiv.

132. Ibid., ch. 7, p. 252 [194].

133. Ibid., ch. 6, p. 208 [162]. Dewey also maintains that survival requires selection by the organism, the basis for alteration of the environment (ibid., ch. 7, p. 256 [197]). This "selective bias," which is value relevant, is again organic, not specifically human.

134. 1922, III, 5, p. 196 [M14].

135. 1925, ch. 5, p. 185 [L1: 145]. Dewey thinks that the best evidence against animal thought is that they do not use tools. However, some animals do use tools.

136. Ibid., Preface, p. xiii.

137. Ibid., ch. 7, p. 272 [208].

138. As Hartshorne and Callicott have noted, humans are not the purpose of evolution, a point on which Darwin has been misinterpreted. The lower is not for the sake of the higher even if the higher has evolved out of the lower. For Callicott's views see *In Defense of the Land Ethic, Essays in Environmental Philosophy* (Albany: State University of New York Press, 1989); for Hartshorne's see "The Rights of the Subhuman World," *Env. Eth.*, I, 1, 1979.

139. 1922, I, 3, p. 47 [M14]. Dewey anticipated Hartshorne and Callicott on this point (see n. 141). On evolution see ibid., IV, 1, p. 263.

140. I will address this point in more detail in chapter five.

141. For Weston's view see "Beyond Intrinsic Value, Pragmatism in Environmental Ethics," *Environmental Pragmatism* (New York: Routledge, 1996) ch. 14, p. 285 (reprinted from *Env. Eth.*, 7/4, 1985), which I discussed in the prologue to chapter two. Weston's view was formulated in dialogue with Katz's.

Chapter Three

All citations of the works of Dewey will be referred to by their initial date of publication. These will be followed by the University of Illinois edition of the Complete Works in brackets followed by the page number of the latter where appropriate. For example, "Middle Works" Vol. 4 page 18 would be [M4: 18]. See *John Dewey, The Middle Works* and *John Dewey, The Later Works*, ed. by Boydston, J. (Carbondale: So. Illinois Univ. Press, 1984). For a list of these works see the beginning of the notes.

1. See the prologue for chapter two of this book and Katz, E., "Searching for Intrinsic Value: Pragmatism and Despair in Environmental Ethics," in Katz and Light, eds., *Environmental Pragmatism* (New York: Routledge, 1996) ch. 15, p. 307 ff. (reprinted from *Environmental Ethics*, Vol. 9, #3, 1987). For the thesis that Dewey undermined or eliminated intrinsic value see Beardsley, M., "Intrinsic Value," *Phil. and Phenomenological Research*, 26, 1965, p. 6 ff.; Baylis, C., "Grading, Values and Choice," *Mind*, 67 (1958), p. 490; Garner, R. T. and Rosen, B., *Moral Philosophy* (New York: Macmillan, 1967) p. 129 (my source on the last two articles is Willard's article below);

and Nissen, B., "John Dewey on Means and Ends," *Philosophy Research Archives*, 3, 1977, pp. 709–738. E. T. Mitchell argues that Dewey pays little attention to intrinsic values ("Dewey's Theory of Valuation," *Ethics*, LV, July 1945, 287–297). For the contrary thesis, which this book supports, see Gouinlock, J., *John Dewey's Philosophy of Value* (New York: Humanities Press, 1972) and Willard, L. D., "Intrinsic Value in Dewey," *Philosophy Research Archives*, I, 1975, pp. 54–77. It is impossible to read *Experience and Nature* and still try to maintain the anti-intrinsic value reading.

2. There are interesting parallels between Dewey and Nietzsche's instrumental approach to knowledge and other areas outside of values proper.

3. For Dewey's critique of Greek philosophy see *The Quest for Certainty, A Study of the Relation of Knowledge and Action* (New York: Capricorn/ Putnam, 1929/1960) [L4], hereafter 1929. Of course, Dewey is thoroughly modern; in his own mind he is almost premodern, since he states that "the genuinely modern has still to be brought into existence," *Reconstruction in Philosophy* (Boston: Beacon, 1920), [M12: 77 ff.] (1972 ed., hereafter 1920) Intro., p. xxxv. He does not accept the Aristotelian notion of natural *telos*. He absorbed and incorporated Darwin, and thus rejects fixed species or essential forms, for example.

4. "The Field of Value," in Lepley, R., ed., *Value, a Cooperative Inquiry* (New York: Columbia Univ. Press, 1949) ch. 3, p. 64 ff. [L16: 343–357] (hereafter 1949), p. 69 [348]. Cf. "Theory of Valuation," from the *International Encyclopedia of Unified Science* (Chicago: Univ. of Chicago Press, Vol. II, #4, 1939) [L13: 189–254] (hereafter 1939), sect. 4, p. 404 [212] where the notion of instrumental and intrinsic value as coextensive with the reflexive relation of "in and for themselves," and "for their own sake" as the starting point for Dewey's own analysis. For Dewey this is connected to the distinction of prizing and appraising: the latter applies to means, the former to ends (p. 405) [213]. However, his critique of the distinction attenuates it, a point that will be argued later.

5. Ibid., (1939), sect. 5, p. 413 [L13: 221].

6. 1922 [M14], P. I, ch. 2, p. 35.

7. 1939, sect. 2, p. 391 [L13: 199]. By reflexive statement I mean the formula of "for its own sake" and the corresponding instrumental relation of "for the sake of."

8. 1922 [M14], P. III, sect. 6, p. 217.

9. Ibid., I, 2, p. 34. Economics does not constitute an exception, although it does not explicitly connect ends and means. While Dewey concedes that economic decisions are "always strictly instrumental" and do not involve a choice of ends, the ends are taken for granted (ibid., II, 5, p. 204). Dewey argues that means dissevered from ends are "impossible," but also that ends without consideration of means are irrational, amoral and philosophically untenable.

10. Ibid., I, 2, p. 35.

11. 1939, sect. 6, p. 422 [L13: 230].

12. Nissen's argument that Dewey "confuses an end or a goal with the plan to achieve that goal" is simply wrong: Dewey carefully distinguishes end-in-view from goal, while noting the ambiguity of 'end,' as well as deliberation from action. (For Nissen's view see *Phil. Res. Arch.*, 3, 1977, 709–738).

13. 1949, p. 65 [L16: 344]. Cf. 1939, sect. 4, p. 409 [L13: 217]. It is notable that Dewey states that *all* processes have an end, not just human activities. This includes natural processes. Human outcomes are a species of natural processes and the human is within nature in this respect.

14. Ibid., (1939).

15. 1922 [M14], III, 6, p. 209. Dewey argues (ibid.) that even the most important consequences are not necessarily the aim of the act. They stimulate action but do not necessarily coincide with its outcome.

16. Ibid., III, 4, p. 195.

17. 1939, sect. 6, p. 425 [L13: 233].

18. 1920, ch. 7, p. 175 [M12: 172 ff.], emphasis mine. This passage is in the context of a defense of value pluralism, and the intrinsic value of a "general" like health may be comparative, i.e. its value cannot be reduced to another *kind* of value. However, Dewey does have a qualified notion of intrinsic value, as I will argue.

19. 1922 [M14], III, 9, p. 251.

20. 1939, sect. 6, p. 423 [L13: 231].

21. For a counterargument see Beardsley, M., "Intrinsic Value," *Phil. and Phenomenological Research*, 26, 1965, p. 6 ff. and the other sources listed in n. 1. Beardsley identifies Dewey as a critic of intrinsic value without specifying how or where Dewey argues against the notion as opposed to refining it. For a view that is supportive of my own, from a different perspective, see Gouinlock, 1972.

22. Incredibly, B. Nissen states that Dewey "fails to invalidate the instrumental value/intrinsic value distinction" (1977, p. 709). Dewey indeed fails to invalidate the distinction, since he upholds it. Nissen concedes that an end may be "consummatory" in ways that means are not. But he argues that Dewey makes no attempt to "integrate" art as consummatory into the means-end analysis. Nissen's argument that the distinction is destroyed because means, like ends, can have intrinsic qualities and ends can be appraised ignores other features of the distinction, which Dewey upholds. Nissen seems unaware of the work of Gouinlock, which addresses precisely the issue of the integration of art as consummatory with means and ends; and Willard (n. 1). Both authors have extensively documented Dewey's notion of intrinsic worth.

23. 1920, ch. 7, p. 171 [M12: 172 ff.].

24. 1922 [M14], III, 9, p. 249.

25. The notion of intrinsic value as a consummation of experience is emphasized particularly in *Experience and Nature*, 1925, [L1] (New York: Dover, 1958 edition of the 1929 revision, hereafter 1925), esp. ch. 10. It should be noted that for Dewey, experience is behavioral not private, thus consummations of experience are not subjective.

26. Ibid., p. 80 [L1: 70].

27. Ibid., ch. 3, p. 115 [96].

28. As Gouinlock has noted, the critique of transcendent ends is not a critique of all ends only those remote from present activity (1972, ch. 5, sect. 2).

29. 1949, p. 67 [L16: 346]. This is also an implicit critique of Moore's analysis of good as a nonnatural predicate.

30. 1925, ch. 3 [L1].

31. 1939, sect. 4, p. 406 [L13: 214].

32. 1922 [M14], III, 6, p. 207. Dewey is of course referring in this passage to ends-in-view, not consequences.

33. 1939, sect. 4, p. 408 [L13: 216]. Cf. 1949, p. 74 [L16: 354]. Nissen (1977) argues that although Dewey is correct in arguing that intrinsic does not mean having no relation to anything else, most of the tradition does not understand intrinsic as out of relation. This criticism is correct, but Dewey's point is aimed at certain figures in the

tradition, e.g., Kant and Plato, not the whole tradition. As Gouinlock has noted (1972, ch. 3, p. 145 ff.), Dewey is denying absolute ends, ends out of any relation.

34. 1922 [M14], III, 8, p. 241. Dewey argues that remote ideals, removed from the immediate situation, do not induce activity effectively. The result is that such ideals are considered unattainable (1925, ch. 10, p. 280 ff. [L1: 214]). Unfortunately in Dewey's view, "popular ideals" are "infected" with the notion of a "fixed end beyond activity" at which we should aim. Such ends in themselves come before aims, if this model is followed, which ought to be aimed at whether they are or not (1922, [M14] III, 6, pp. 207–208).

35. Ibid., [1922, M14] p. 211.

36. 1939, sect. 6, p. 424 [L13: 232].

37. 1920, ch. 7, p. 162 [M12: 172 ff.]. Dewey relates this search to the quest for certainty, i.e. to a certain good or end. This quest seeks security, "a demand for guarantees in advance of action" (1922, [M14] III, 6, p. 219).

38. Ibid., (1922 [M14]) p. 215. The doctrine of fixed ends has rendered men "careless" in their "inspection of existing conditions" (ibid.). The result has been that they are thrown back on past habits, and become pessimistic about changing their circumstances.

39. Ibid., pp. 214–215.

40. Ibid., p. 212.

41. 1939, sect. 6, p. 422 [L13: 230].

42. I will examine the elements of deliberation in detail in the next chapter, "Dewey's Holism."

43. 1925, ch. 10, p. 259 [L1: 198]. Cf. White's critique of Dewey in "Value and Obligation in Dewey and Lewis," *Phil. Rev.,* 58, July 1949, 321–329, in which Dewey is charged with deriving an ought from an is, desirable from desired, enjoyable from enjoyed. This is clearly not true for Dewey, since he criticizes Mill for the same move, argues against an empirical, "given" or impulsive view of 'desire' (cf. the next chapter on the relation of considered desire to mere impulse), and does not equate desirable with obligation. For a critique of White's argument from three other perspectives see Gouinlock 1972, ch. 3, Hook, S., "The Ethical Theory of John Dewey," pp. 49–70 of *The Quest for Being and Other Studies in Naturalism and Humanism* (New York: St. Martin's Press, 1961), and Hurley, Paul E., "Dewey on Desires: The Lost Argument," *Trans. of the Charles S. Peirce Society,* 24:3, 1988, 509–519. (Willard, op. cit., 1975 argues that Dewey did not deal "adequately" with the distinction of desire and desirable.) Cf. n. 83 and 86.

44. Gouinlock has argued further that "values . . . are not equivalent to ends-in-view" (1972, ch. 3, p. 130). (However, I do not agree with Gouinlock that value is confined to the consummatory phase of events.)

45. 1939, sect. 7, p. 433 [L13: 241]. The means to end relation closely follows the cause to effect relation, as it does for the tradition. Thus as Gouinlock has noted (1972, ch. 5), it is dependent on facts about nature.

46. Ibid., sect. 4, p. 405 [L13: 213].

47. 1922, [M14] I, 2, p. 34.

48. Dewey provides an example in the form of the separation of a product and its consumption. This is exhibited especially in "accidental consumption and extravagance," what Veblen would have called "conspicuous consumption," which Dewey contrasts with "normal consummation or fulfillment of activity" (1922, [M14] III, 9; for

Veblen's view see his *The Theory of the Leisure Class*). In this normative notion of natural consumption by contrast with socially arbitrary forms, Dewey anticipates the Critical Philosophy of the Frankfurt School.

49. 1939, sect. 4, p. 407 [L13: 215].

50. 1949, p. 70 [L16: 350].

51. 1922, [M14] III, 6, p. 211. Cf. 1949, p. 72 ff. [L16: 352 ff.] where the same criticism is made in the late period.

52. Ibid., (1922 [M14]), p. 212.

53. 1939, sect. 5, p. 416 [L13: 224].

54. Ibid., sect. 6, p. 422 [230]. Nissen's argument (1977, p. 709) that Dewey fails to show "that the relation of means to ends is symmetrical" and "that means necessarily resemble their ends," ignores the entire tradition, based on Aristotle, in which ends can later be used as means. A hierarchy of ends is crucial to Aristotle's arguments.

55. Ibid., p. 429 [237].

56. Ibid., p. 423 [231].

57. 1922, [M14] I, 2, p. 35. The continuity of means and ends is also a doctrine maintained from the early to the late period: cf. 1949, p. 72 ff. [L16: 352 ff.].

58. 1939, sect. 6, p. 421 [L13: 229].

59. Ibid., sect. 5, p. 418 [226] emphasis in original.

60. 1908/1932, ch. 11, p. 198 [L7: 184].

61. In turn, growth is tied to self-fulfillment, evidence for Sahakian's contention that Dewey's ethic is an ethic of self-development or self-realization. To grow is to develop; fulfillment is the culmination of growth. However, Dewey argues that "continuity of growth" is the "alternative to fixity of principles and aims" (1922, [M14] III, 9, p. 251). Growth must not stop if any one end is achieved, but continue. Nor is growth future oriented, which an end as goal might imply. Dewey urges that we prize opportunities for present growth.

Self-realization must be in harmony with other considerations. As Pappas has argued, Dewey's is not a new version of character or agent-centered ethics. For Dewey is seeking a unity of the moral self and its acts. (For Pappas view see Pappas, G. F., "To Be or to Do, John Dewey and the Great Divide in Ethics," *Hist. Phil. Quart.*, 14:4, 1997, 447–472. For Sahakian's view see *Ethics: An Introduction to Theories and Problems*, New York, Barnes and Noble Books, 1974.)

62. This is probably the only "Platonic" element in Dewey's ethics. Cf. *Gorgias*, 303 ff. and *Philebus* where Plato describes good as residing in the "order and harmony of the parts."

63. 1920, ch. 7, p. 177 [M12: 172 ff.].

64. In the tradition, intrinsic value and the highest good have often been equated, but they are distinct. While the highest good requires a notion of intrinsic value, the reverse is not true, for there can be a multitude of intrinsically valuable items without any one of them forming the highest one.

65. Ibid. (1920), [M12: 172 ff.]

66. Ibid. p. 166.

67. Dewey's theory of nature and why goods can be considered natural was discussed in ch. 2. In this context it can be stated that they are natural in the sense of naturally occurring (e.g., health) and positive directions of change or objectives whose pursuit is instrumental to such a change.

68. 1922, [M14] III, 6, p. 209.

69. Ibid., p. 210.

70. Cf. 1939, sect. 8, p. 443 [L13: 421], where he calls for a theory of value as an "effective instrumentality."

71. 1929, ch. 1, p. 35 [L4: 28]. Cf. 1922, [M14] III, 1, p. 165 where Aristotle is viewed as the source of the idea that good is a perfection and attribute of being; static perfection is the highest ideal or excellence, which "excludes all potentiality." This identifies good as an attribute of fixed being, a perfection, with an ideal, which as beyond change is fixed. Thus the origin of the idealist theory of value lies in Aristotelianism for Dewey.

72. Ibid., (1922 [M14]) IV, 1, p. 284.

73. Ibid., III, 9, p. 244.

74. Thus Dewey commends the Epicurean emphasis on present goods, although not its ethic of withdrawal (ibid., IV, 1, p. 268 ff.).

75. 1922, [M14] IV, 2, p. 271. Thus the origin of the so-called distinction of is and ought lies in idealism, despite Hume's articulation of it.

76. Ibid., III, 6, p. 218.

77. 1939, sect. 7, p. 437 [L13: 245]. I would argue that this is precisely the opposite of Kant's view, i.e., that control of desires by reason in the form of moral norms regulating formation of ends of desire is the problematic of Kantian ethics. Kant certainly believes that control of desires is feasible, and is necessary for moral worth. However, Dewey's point that Kant separates supreme value from situations is valid.

78. 1922, [M14] III, 6, p. 214.

79. Ibid., I, 2, p. 28. Again, I would argue that Kant does not argue this, for maxims of practice or practical imperatives are often based on experience.

80. Ibid., I, 3, p. 49.

81. Ibid., III, 8, p. 231. Dewey accuses the Utilitarians of sacrificing creativity and other more worthwhile values for "bourgeois comfort" and acquisition. Commercialism is promoted over social service.

82. The many factors involved in moral deliberation and justification will be treated in chapter four. Dewey's criticisms of the interest theory will be discussed in more detail in a projected future work, *Pragmatic Consequentialism*.

83. 1939, sect. 7, p. 434 [L13: 242]. Cf. Sellars on the "myth of the given."

84. See 1949 sect. 1 [L16]. I will argue this point in more detail in my book in progress, *Pragmatic Consequentialism*.

85. This point is argued in Meta. IV; Cooper claims that Aristotle believed that there is deliberation about ends in ethics. See Cooper, J., *Reason and Human Good in Aristotle* (Cambridge: Harvard Univ. Press, 1975) P. I.

86. 1939, sect. 7, p. 434 [L13: 242]. This answers White's article (1949, op. cit.; cf. n. 43) on the derivation of an ought from an is. A desire for Dewey is not given or an 'is' simple (cf. text quoted in n. 83). It is conceptually and rationally mediated, as well as evaluated in relation both to other desires and to the whole series of factors in holistic consideration. As Gouinlock has noted, desire has "ideational" elements (1972, ch. 7, p. 280), which may be determined by intellect. Intellect is not a "slave of the passions" for Dewey but can "reconstitute" desire. Caspary ("Judgements of Value in John Dewey's Theory of Ethics," *Educ. Theory*, 40:2, 1990, 155–169) has noted that the distinction of valuing and valuation parallels the distinction of desired-desirable. He

quotes where Dewey states that "judgements about values . . . about that which should regulate the formation of our desires, affections and enjoyments" (1929, p. 265 [L4: 212). In other words, desires are formed in terms of values but they are not the basis of values.

87. This may reflect the influence of the Hegelian dialectic.

88. 1939, sect. 7, pp. 431–432 [L13: 239–240]. Dewey notes that ideals when brought into existence are existent not ideal, i.e., not outside the process (1925, ch 2, p. 62 [L1: 58]).

89. 1922 [M14], I, 2, p. 27. The effectiveness of habits is increased by the use of intelligence, as I noted earlier. In the face of new situations there may be a need for "revision, development and readjustment." This will result in new standards that in turn increase ends (1920, ch. 7, p. 175 [M12: 172 ff.]).

90. 1925, ch. 7, p. 269 [L1: 206].

91. 1922 [M14], III, 6, p. 209 as well as numerous texts in *Experience and Nature*. The relation of value to Dewey's naturalism has been extensively analyzed by Gouinlock, 1972.

92. For a concurring view, see Gouinlock, 1972, ch. 1, sect. 1. Gouinlock notes that dualism is entrenched due to the denial of qualities in nature (ch. 2, sect. 2). For an argument that good is "subjective" in Dewey, see Honneth, A., "Between Proceduralism and Teleology, an Unresolved Conflict in Dewey's Moral Theory," *Trans. of the Charles S. Peirce Society*, 34:3, 1998, 689–711, p. 701. Dewey explicitly repudiates subjectivity in philosophy, including in value theory, in *Experience and Nature*.

93. 1949, ch. 3, p. 66 [L16: 345].

Chapter Four

All citations of the works of Dewey will be referred to by their initial date of publication. (For a list of these works see the beginning of the notes.) These will be followed by the University of Illinois edition of the Complete Works in brackets followed by the page number of the latter where appropriate. For example, "Middle Works" Vol. 4 page 18 would be [M4: 18]. See *John Dewey, The Middle Works* and *John Dewey, The Later Works*, ed. by Boydston, J. (Carbondale: So. Illinois Univ. Press, 1984).

1. Dewey, *Reconstruction in Philosophy* 1920, [M5] (Boston: Beacon, 1972 ed., hereafter **1920**) Introduction. Cf. "Experience, Knowledge and Value, a Rejoinder," in Schilpp, P., ed., *The Philosophy of John Dewey* (New York: Tudor Publ., 1951) where Dewey calls for moral "construction."

2. *Human Nature and Conduct* 1922, [M14] (New York: Modern Library, 1957, hereafter **1922**) I, 3, p. 53. Holmes has argued that Dewey in his ethics comes down more on the side of "personal choice," rather than "conventional moral rules." (See Holmes, R. L., "John Dewey's Moral Philosophy in Contemporary Perspective," *Rev. Meta.*, XX: 1, 77, 1966, 42–70.) This is at best an exaggeration, as a glance at either *Ethics* (Dewey, John, *Ethics*, New York: Henry Holt, 1908, 1932 revision, [M5/L7] co-authored with J. Tufts, hereafter **1908**) ch. 10 or, 1922, [M14] Part IV, would confirm. In the former, social customs are seen as the source of moral rules and sufficient unless

in conflict or if a new, problematic situation arises. (For the relation of moral reflection to custom see 1908, esp. ch. 10, sect. 1 and 5 [L7]). As Gouinlock remarked, "morally problematic situations are social in nature" ("Dewey's Theory of Moral Deliberation," *Ethics*, 88:3, 1978, pp. 218–228, p. 225. Cf. his *John Dewey's Philosophy of Value* (New York.: Humanities Press, 1972) ch. 5, (a point echoed in the chapter he wrote on "Dewey," ch. 10 of *Ethics in the History of Western Philosophy*, ed. Cavalier, R., Gouinlock, J. and Sterba, J. P. (New York: St. Martin's Press, 1989) pp. 306–332, sect. 4 where moral deliberation is described as "social or democratic"). Morality is conceived of by Dewey as "primarily social" in the latter work (1922 [M14]) and throughout Dewey's ethical writings (see also below on the different kinds of consequences that are morally considerable, of which social consequences are primary). Morals are the condition of society for Dewey. The elements of self-development and self-realization, while present, are conditioned by a host of regulative elements, including moral principles as instruments, intelligence, growth, and character. Self-development is not a matter of personal choice, as what is realized is character, a standard of a sort that is trans-personal. Gouinlock has also pointed out that, in Dewey's view, society shapes the individual through customs, culture, the inculcation of habits and character, and other social forces (1972, ch. 2, sect. 3). The atomic individual is not a "given," for Dewey, but a creation of individualistic society. Thus Bowers' contention that Dewey would have rejected Solzhenitsyn's definition of freedom as "self-restriction," and that he emphasized "individuals making up their own minds and selecting their own values" ignores the priority of society in moral deliberation for Dewey. For Bowers' view see *Education, Cultural Myths and the Ecological Crisis* (Albany: State Univ. of New York Press, 1993) ch. 2, pp. 91 and 104. Cf. n. 45 and 58.

3. 1908, ch. 10, p. 171–2 [L7: 162–3].

4. 1920, "Introduction," et al. [M12].

5. *The Quest for Certainty, A Study of the Relation of Knowledge and Action*, 1929, (New York: Capricorn/ Putnam, 1960, hereafter **1929**) ch. 1, p. 32 [L4: 26].

6. 1922, [M14] IV, 1, pp. 257–258. This scope is limited by acts done from habit or impulse, thus not all acts are moral in the strict sense. However, even habitual and impulsive acts are morally deliberable or considerable.

7. 1920, ch. 7, p. 177 ff. [M12: 172 ff.]. The phrase "meliorist" has probably been kidnapped from William James.

8. 1908, p. 364 ff. [L7: 328 ff.] H. Stuart argued that Dewey did not adequately distinguish the moral method from the experimental method; moral deliberation from thought processes in general. (See "Dewey's Ethical Theory," in *The Philosophy of John Dewey*, ed. Schilpp, P., New York: Tudor Publ., 1939/1951, pp. 293–333.) While there is a close parallel between the experimental and moral methods, since Dewey was concerned to modernize ethics, there are distinct differences. The experimental method "in general" is not concerned with better or worse alternatives, social and individual consequences, or valuations. Morality is concerned with future practical outcomes; the experimental method with practical outcomes only ultimately. Dewey was concerned to expand the sphere of the morally considerable, however, to include, e.g., the economic.

9. Pepper has defined Dewey as a "contextualist" in Fern, V., ed., *A History of Philosophical Systems* (New York: Philosophical Library, 1950) et al. and other pragmatists influenced by Dewey have adopted this label (Lepley, 1949), although Dewey

himself does not adopt it. Dewey is a contextualist insofar as existing conditions or the present situation is a vital element in defining the problem for value judgment, but not if a situation ethic is meant. Past experience in the form of principles are brought to bear as instruments for resolution of such problems. This point will be covered later.

10. 1908 [L7]. Dogma is the appeal to authority, including the supernatural or tradition, over evidence.

11. Ibid.

12. For parallel analyses of the elements involved in moral deliberation see Gouinlock, 1972, ch. 3; and Caspary, W., 1990. My list differs in a few details from these authors. Gouinlock notes that every situation has certain elements, as part of a process, including a beginning and an end, and a cause and effect (ibid., ch. 2). For Caspary's view see Caspary, W., "Judgments of Value in John Dewey's Theory of Ethics," *Educational Theory*, 40/2, 1990, pp. 155–169.

13. "Theory of Valuation," from the *International Encyclopedia of Unified Science* (Chicago: Univ. of Chicago Press, Vol. II, #4, 1939, hereafter **1939**) sect. 5, p. 414 [L13: 222].

14. 1922 [M14], III, 7, p. 221.

15. 1908, ch. 10, sect. 2, p. 178 [L7: 168].

16. Unlike more psychological or subjective theories of desire, then, e.g., Descartes', Dewey agrees with the more conative theory of desire pioneered by Perry and Prall. Desire includes a striving or urge to act, not a mere feeling of lack or a wish.

17. 1922, [M14] I, 1, p. 24.

18. 1939, sect. 5, p. 413 [L13: 221].

19. 1922, [M14] IV, 1, p. 264.

20. Social conditions and "cultural conditions and institutions" are also overlooked by the interest theory, which thus neglects the way desires and the values that satisfy them are implanted by a culture (1939, sect. 8, p. 444 [L13: 252]). Desires are not entirely autonomous in a subject but may reflect a social situation by which they have been formed. This quote also partly answers Bowers' contention that Dewey ignored culture (for Bowers see n. 1). Culture shapes desires, which are normatively regulated by cultural taboos and prizings.

21. Ibid.

22. Ibid., p. 432 [240]. Nissen's argument that Dewey confuses ends and plans to achieve ends ignores the clear distinction of ends-in-view from ends in the sense of consequences made by Dewey. (For Nissen's view see "John Dewey on Means and Ends," *Phil. Res. Archives*, 3, 1977, 709–738). His argument that for Dewey we must "always appraise an end in terms of the means on which it rests" (ibid., p. 729) is true, but ignores the relation of means and ends to other elements in the problematic situation.

23. Dewey is careful to separate the end as a goal from the end as a consequence, an ambiguity that is likely to cause confusion (1939, sect. 6, p. 428 [L13: 236]). See previous note.

24. Ibid., (1939) sect. 5, p. 414 [222].

25. Ibid., p. 432 [240].

26. 1929, ch. 10, p. 260 [L4: 208]. He makes a similar distinction between the enjoyed and enjoyable, the satisfying and satisfactory. Thus the view proposed by E.T. Mitchell that Dewey believed that "choice is ultimately impulsive," is untenable. (For

Mitchell's view see Dewey's Theory of Valuation," *Ethics*, July 1945, LV, pp. 287–297). Cf. below on the role of intelligence in evaluation and choice.

27. 1939, sect. 7, p. 433 [L13: 241]. Dewey believed that the interest theory overlooked this element in the formation of desires.

28. Ibid., sect. 3, p. 396 [L13: 204].

29. Ibid.

30. 1920, ch. 7, p. 163 [M12: 172 ff.]. Dewey argues that choices must be made in all ages of life. This is an implicit critique of Aristotle, who thought the formative choices in life must be made early. Cf. "plans of life" in Rawls, *A Theory of Justice*.

31. Ibid., p. 164.

32. Gouinlock in 1972, ch. 7 particularly stresses the central role of intelligence in Dewey's concept of "value in nature."

33. A further role of intelligence is in relation to the meaning of activity: intelligence creates the connections that constitute meaning for Dewey. "An activity has meaning in the degree in which it establishes and acknowledges variety and intimacy of connections" (1922, [M14] IV, 1, p. 270). Intelligence again has a role to play in the study of methods of acting to enhance values. "The chief consideration in achieving concrete security of values lies in the perfecting of *methods* of action" (ibid., I, 2, p. 36).

34. *Experience and Nature* 1925 (New York: Dover, 1958 ed. based on the 1929 revision) hereafter **1925**, ch. 3, p. 108 [L1: 90].

35. 1922, [M14] III, 5, p. 197.

36. "Intrinsic Value in Dewey," *Phil. Res. Archives*, I, 1975, 54–77, pp. 64–65.

37. 1908, p. 215 [L7: 199–200].

38. 1929, ch. 1.

39. 1939, sect. 7, p. 433 [L13: 241]. Cf. 1922, [M14] IV, 2: "the quality of these consequences determines the question of better and worse." Thus Dewey consistently maintains this position from earlier to later works.

40. "The attained end or consequence is always an organization of activities . . . coordination of activities . . ." (ibid., (1939) sect. 6, p. 428 [236]).

41. 1922, [M14] IV, 1, pp. 265–266.

42. Ibid., III, 4, p. 249. Many of these goals have a Platonic flavor, e.g., recovering harmony. However, harmony is more dynamic for Dewey than for Plato.

43. 1939, sect. 1, p. 382 [L13: 190]. In a sense, pragmatism in general is an extension of moral consequentialism to all of philosophy.

44. Ibid., sect. 6, p. 422 [230].

45. As Gouinlock has noted (1972, ch. 6, sect. 2), there is no dichotomy of individual and social good in Dewey. The latter would be egoistic and based on a false psychology of the isolated subject. Thus despite elements of self-realizationism in his ethics, the self is not foundational in either the moral or epistemological sense. Cf. Pappas' argument that Dewey attempted to overcome the character/action ethical dichotomy. (For Pappas' view see Pappas, G. F., "To Be or to Do, John Dewey and the Great Divide in Ethics," *Hist. Phil. Quart.*, 14:4, 1997, 447–472.) Cf. n. 1.

46. Dewey argues that morals should be limited to "foreseen and desired consequences" in 1908, ch. 10, sect. 4, p. 184 [L7: 173]. Cf. 1925, ch. 1, p. 21 and ch. 2, p. 43 [L1: 28 & 44]. This does not differ significantly from Moore's view on consideration of consequences and the problem of unintended consequences.

47. 1922, [M14] I, 3, p 46. Dewey also argues that morals should not be too closely identified with consequences either, in the same section.
48. Ibid., IV, 1, p. 259.
49. Ibid.
50. Ibid., III, 5, p. 75. A limit to consequences is that acts must be in our control, thus "still to be performed" (ibid., I, 1, p. 21). This answers Altman's contention that if murder were to enhance growth, it might be justified in Dewey's ethics. Altman's argument also ignores the holistic relations to which growth as an overall end is part. Growth is not foundational for Dewey, a point that will be discussed in the next section. For Altman's view see "John Dewey and Contemporary Normative Ethics," *Metaphilosophy*, 13/2, 1982, 149–60, p. 158.
51. Honneth argues both that for Dewey there is no universal obligation to respect human beings and that Dewey brings in a "social element" by arguing that the question of what is morally right is presumed in every rational process of deliberation (Honneth, p. 702). These arguments contradict one another, since if a social element is presumed in every process of deliberation, it is obligatory, and as involved in every such process is universal. (For Honneth's view see "Between Proceduralism and Teleology, an Unresolved Conflict in Dewey's Moral Theory," *Transactions of the Charles S. Peirce Society*, 34:3, 1998, pp. 689–711.)
52. 1908, p. 387 [L7: 348]. Growth and "self-fulfillment" are equated in 1922, [M14] III, 9, p. 251.
53. 1920, ch. 7, p. 186 [M12: 172 ff.].
54. 1908 p. 384 ff. [L7: 346]. This involves a relation of the whole (society) to its members and of the members (part) to the whole.
55. In 1908, ch. 10, p. 184 [L7: 173] consequentialist ethics is rejected as "one-sided."
56. 1922, [M14] III, 8, p. 231. Another holistic formulation is conduct described as binding together acts into a whole (1908, ch. 10, p. 181 [L7: 170]).
57. The use of organic models may also reflect the influence of Hegel and Darwin.
58. 1939, sect. 8, p. 444 [L13: 252]. Dewey argues the point that subjective ethics is anti-social in 1922, [M14] IV, 1, p. 269.
59. Cf. 1920, ch. 7, p. 169 [M12: 172 ff.] where the "intellectual" value of drawing general classes of goods from comparable cases is mentioned to 1939, sect. 6, p. 424 [L13: 232], where general ideas of value are used as intellectual instrumentalities. They have no bearer except as an idea, so are not cases of fixed value. Nevertheless, they have instrumental value as qualities.
60. Pappas (1997, p. 447 & 452 see n. 45) also characterizes this divide as the "following of rules" as opposed to the "cultivation of traits." For a counterargument see Honneth's article (n. 51).
61. By attenuating the distinction of ends and means, Dewey argued convincingly for the instrumental role of ends. Ends, principles and past experiences are instrumental in the unique value situation. More, Dewey revalued means as worthy of greater consideration, an extension of value, since means are not considered subordinate in value or of inferior worth in his analysis. I have argued this point in more detail in the last chapter, "Dewey's Instrumentalism."
62. See ch. 1. For Katz' view, see "Searching for Intrinsic Value: Pragmatism and Despair in Environmental Ethics," reprinted in Katz and Light, eds., *Environmental*

Pragmatism (New York: Routledge 1996) ch. 15, p. 307 ff. Reprinted from *Environmental Ethics*, Vol. 9, #3, 1987.

63. I will briefly treat the other pragmatists in the Epilogue. Intrinsic value in some other pragmatists will be documented more fully in a book in progress, "Pragmatic Consequentialism," which will incorporate the value theories and moral approaches of C. I. Lewis and F. C. S. Schiller.

64. This is despite foundational elements and occasional expressions that seem to point away from holism, e.g., his use of experience and empirical, as in British empiricism; induction from circumstances to generality, etc. In general, these are incorporated into his more holistic analysis. Dewey argues that his nonfoundational model of justification does not involve an infinite regress (1939, sect. 6, p. 425 ff. [L13: 233 ff.]), an objection that can be traced back to Aristotle. For the circle of justification is finite, not infinite and is ultimately defined with reference to the situation.

65. Dewey has often been misread as doing away with intrinsic value, e.g., by Beardsley, M., "Intrinsic Value," *Phil. and Phenomenological Research*, 26, 1965, p. 6 ff.; and Nissen, B., "John Dewey on Means and Ends," *Philosophy Research Archives*, 3, 1977, pp. 709–738, inter alia. Gouinlock has gone to some length to document the error of this view (in 1972, passim.), which has been a persistent myth and hard to kill. One reason readers may have been mislead about Dewey's treatment of intrinsic value is his holism. Intrinsic value is morally considerable but not foundational for Dewey. No one factor can be the whole warrant in moral decisions without consideration of all. Thus Dewey's holism is both a rejection of foundationalism in ethics, and thereby consequentialism of the Utilitarian type, since the former is closely connected with the latter.

66. 1939, sect. 8, p. 440 [L13: 248].

Chapter Five

I will cite works of Dewey, which will be referred to by their initial date of publication. (For a list of these works see the beginning of the Notes.) These will be followed by the University of Illinois edition of the Complete Works in brackets followed by the page number of the latter where appropriate. For example, "Middle Works" Vol. 4 page 18 would be [M4: 18]. See *John Dewey, The Middle Works* and *John Dewey, The Later Works*, ed. by Boydston, J. (Carbondale: So. Illinois Univ. Press, 1984).

1. Callicott, J. Baird, *In Defense of the Land Ethic, Essays in Environmental Philosophy* (Albany: State University of New York Press, 1989) passim.; Rolston, H., *Environmental Ethics, Duties to and Values in the Natural World* (Philadelphia: Temple Univ. Press, 1988). See chapter one.

2. As I noted in the last chapter, there are parallel analyses of the elements involved in moral deliberation in Gouinlock, *John Dewey's Philosophy of Value* (New York.: Humanities Press, 1972), ch. 3; and Caspary, W., "Judgments of Value in John Dewey's Theory of Ethics," *Educational Theory*, 40/2, 1990, pp. 155–169. My list differs in a few details from these authors. Gouinlock notes that every situation has certain elements, as part of a process, including a beginning and an end, and a cause and effect (ibid., ch. 2).

3. The major is that moral considerability is based on intrinsic value. Cf. n. 5.

4. See above, n. 1. Callicott (ibid.) is of two minds on this issue, since he realizes that the land ethic involves a holistic view but is reluctant to give up his Humean model of value individualism based on moral sentiments. Dewey provides a consistent holistic model influenced, as were Callicott and Leopold, by Darwin.

5. The model was even adopted by non-utilitarians such as Rolston. But can duty be based on value? There is no logical basis: one cannot deduce obligations from value, a point Kant noticed. The alternative is to define obligation as intrinsically valuable, the strategy of Perry and Prall; or to define it solely in terms of intrinsic value, the strategy of Mill and Moore. A further question is whether morals should conform to a basically ontological model (Aristotle) rather than one of agency as in Dewey. Dewey's view represents a certain amount of liberation of morals from ontology.

6. Thus culling wild herds of deer may be justified for the sake of a habitat or the biosphere as a whole.

7. This point is made in 1939, sect. 6, p. 427 [L13: 235].

8. 1922, [M14] III, 4, p. 194. Such expansiveness is a norm itself: an obligation to expand one's perspective. This is a subsidiary argument for extension of moral consideration to animals.

9. The arbitrariness of subjective intrinsic value theories, e.g., the idea that value is what is desired—value is to be desired—is strongly criticized in 1939 [L13], throughout. Not only are such theories narrow, they rule out intelligent consideration of ends, means, and other important factors. Impulse replaces intelligent action and justifies nefarious means. Dewey argued for the role of intelligence in critically evaluating desires. Thus the contention that Dewey himself somehow has a conative theory of value is a misreading.

10. Notably Rachel Carlson. Bowers criticizes Dewey for overemphasizing the role of science in solving problematic situations. As I noted in the Prologue to chapter two, however, the experimental method, adopted from science, is used only as a last resort, when traditional methods have not worked. Bowers ignores Dewey's view of science as problem solving. In other words if environmental degradation is a problem, science may be a tool for dealing with the problem and formulating solutions. This does not exclude traditional views, however. On the contrary, if traditional ways of living are instrumental to living well within an environment, they may be part of a solution or even morally required. For Bowers view see *Education, Cultural Myths and the Ecological Crisis* (Albany: SUNY Press, 1993), ch. 2. Science also, contrary to Bowers, "gives voice" to trees, deer and mountains since it establishes their biological and environmental conditions.

11. Dewey's view of nature and of human nature as within a natural environment is covered in chapter two, "Dewey's Naturalism."

12. 1922, [M14] Introduction, p. 3; cf. section IV. Cf. valuations as distinctly connected to human behavior in 1939, sect. 8, p. 446 [L13: 254]. Valuation is, however, part of the process of valuing activity as a whole; value can attach to the non-human and even the nonliving. The field of value events is "life processes of selection-rejection" i.e., life, not just human life (1949, sect. I [M16, p. 343]).

13. E.g., the substitution of conventional respectability for character. See 1922, [M14] Introduction, p. 5 ff. The cut off of morals from human nature also cuts off the body from moral consideration.

14. Ibid., p. 13.
15. Ibid., p. 6. Cf. p. 9.
16. Ibid., I, 1, p. 19. These include the social environment.
17. Ibid., I, 6, p. 78 ff. Dewey claims that acts should be analyzed in terms of habits and habits in terms of "education, environment and prior acts" (ibid., I, 3, p. 44). Further, the development of habits takes place in an environment, i.e., activity in a setting. Dewey mentions the "objective conditions in which habits are formed and operate" (ibid., I, 4, p. 65). Thought requires independent "organic structures," which in turn require an environment (1925, ch. 7, p. 285 and 290 [L1: 217 & 221]). Intellectual habits also "demand" or require the support of the environment.
18. 1922, [M14] I, 3, p. 50.
19. 1939, sect. 8, p. 437 [L13: 245].
20. 1922, [M14] I, 1, p. 18, emphasis added. Thus Bowers' reading of Dewey's theory of nature, in which it is only "foreground," ignores texts of Dewey's in which the claims of nature and the environment are autonomous from human use and have their own claims. (For Bowers see n. 10).
21. 1925, ch. 8, p. 343 [L1: 258].
22. Ibid., ch. 10, p. 414 [L1: 310]. Dewey was aware of the disruptions of technology in modern life and its affect on separating satisfaction from work; he is less the "poet of technology" than Rorty makes him out to be. He is implicitly critical of the extreme division of labor in the modern factory, of the separation of means from ends in work that it entails, and of "Taylorism" as the tactic of wealth (1922, [M14] III, 9). Though he rejects a return to nature as unfeasible, and as an unintelligent solution to the problems of industrial civilization, he also notes that problems increase in modern civilization despite its solutions to certain ancient problems (ibid., IV, 1, p. 264).
23. Dewey argues in 1922 [M14] that it is the place of religion to show us our place in the whole. This is supportive of stewardship arguments in religious thought.
24. The question is whether it is wise to rest arguments for moral considerability on a "speciesist" model, i.e., one derived from anthropocentric ethics, rather than Dewey's naturalist model. See n. 4 and 5.
25. Value as a trait of nature is covered in 1925, esp. ch. 1 and 10 [L1]. Gouinlock, 1972, gives an extended treatment of this topic.
26. 1908, p. 385 [L7: 347]. Cf. 1920, [M12] ch. 7 where the equal moral importance of every case is argued.
27. 1920, ch. 7, p. 166 [M12: 172 ff.].
28. 1908, p. 383 [L7: 345].
29. Norton, B., "The Constancy of Leopold's Land Ethic," in Katz and Light *Environmental Pragmatism* 1996, ch. 5; and *Why Preserve Natural Variety?* (Princeton: Princeton Univ. Press 1987). See chapter one.
30. Dewey argues that a "change in the environment required to satisfy needs" has no meaning to an animal on its own account. This point is somewhat obscure, but in any case, a change in specific environments can doom a creature, e.g., a tropical species transported to the arctic, a wet grower to the desert, etc. I am not sure if this addresses Dewey's point, however.
31. In an article defending the use of animals in scientific experiments, ("The Ethics of Animal Experimentation," *Atlantic Monthly*, 138, Sept. 1926, pp. 343–346; reprinted in *John Dewey, The Later Works*, ed. by Boydston, J., Carbondale: So. Ill. Univ.

Press, 1984, pp. 98–103), Dewey rejects the idea that animal lives are equal to human lives with some passion. Humans are part of a society with obligations that transcend the merely physical level of sentience of animals. A natural hierarchy is implied in which obligations to human individuals are highest. The emergent qualities of nature are not merely of scientific interest, it would seem, but of moral purport.

32. See n. 31.

33. This issue has been treated in depth by Hans Jonas in *The Imperative of Responsibility, In Search of an Ethics for the Technological Age* (Chicago: Univ. of Chicago Press, 1984).

34. Cf. Norton's arguments for "transformative values" in nature in op. cit., 1984 (see n. 29). Nature may provide unique lessons, which aid in self-development.

35. Since one moral measure is the impact of an arrangement upon habits (1922, [M14] IV, 1, p. 270), wanton destruction might result in the creation of a bad character. (This is to speak nothing of the impact upon animal habits!)

36. Rolston may be accused of this since he argues for the intrinsic value of individuals and then argues that this value is "captured" by predators. For Rolston's view see *Environmental Ethics, Duties to and Values in the Natural World* (Philadelphia: Temple Univ. Press, 1988). Dewey anticipates Rolston's view on holism in many respects.

Epilogue

1. Cartesian residues are (1) subjectivity as autonomous, (2) humans as privileged, while animals are mere mechanisms, (3) the transcendence of nature by the autonomous subject, and (4) the egoism in ethics that follows from an atomic, solipsistic subject.

2. For a study of the normative sciences in Peirce see Potter, V. J., *Charles S. Pierce on Norms and Ideals* (Amherst: University of Massachusetts Press, 1967).

3. I am indebted to Prof. Robert Neville on this point. See *The Highroad Around Modernism* (Albany: SUNY Press, 1992), ch. 1.

4. *An Analysis of Knowledge and Valuation*, The Paul Carus Lectures Series 7 (La Salle, IL: Open Court, 1946, 1962 ed.), ch. 13, sect. 8, p. 412.

5. Ibid., ch. 12, sect. 5, p. 375.

6. Lewis's critique of emotivism is the underlying theme of almost all of the essays in *Values and Imperatives*, ed. Lange, J. (Stanford: Stanford University Press, 1969).

7. 1946, ch. 12, sect. 10 ff., p. 390 ff. Intrinsic can be defined as the contrary of extrinsic, which corresponds to the distinction of means and ends, "for the sake of" and "for its own sake." It can also be defined as intrinsic to an object, i.e., as a property or character of an object: a quality of a being. These senses are often the same for the tradition, particularly Aristotle, but Lewis distinguishes them. The sense of value as a property is 'objective,' of for its own sake 'inherent' and the contrary of extrinsic is intrinsic, which is located in the experiencing subject.

8. Ibid., sect. 12, p. 396. The problem of the objectivity of values arises mostly for a representational theory of knowledge. If qualities are representations, can they inhere in objects? Are they real? Lewis solves this problem by noting that value qualities conduce to satisfactions in a subject, just as other qualities can be predicted to have

certain effects on subjects. If the value relation is like that of cause and effect, i.e., the object causing some effect in the subject, then the object must have an objective value property.

Lewis also solves the epistemological problem of objective qualities in a semi-Kantian manner. Beauty may require the experiencing subject, but if the cause is in the object, and this is universally true, then it is objective.

9. Ibid., ch. 16, sect. 2, p. 483.

10. For Lewis' holism see ibid., sect. 1, p. 479 and sect. 2, p. 483. For esthetic intrinsic value as holistic see ch. 15, sect. 7, p. 478. In the latter section, Lewis approaches Moore's holism.

11. Ibid., ch. 12, sect. 12, p. 395.

12. Ibid., ch. 16, p. 483. The value of the immediate cannot be denied but is not final. Intrinsic value can be "instrumental" or not to "life in general." It must be judged in relation to life as a whole (ibid., p. 479).

13. Also, Lewis begins with analysis of language, not the subject of experience. This provides a link to analytic philosophy and an alternative approach within pragmatism.

14. Ibid., Introduction.

15. Ibid., ch 13, sect. 2.

16. Ibid., ch. 14, sect 6, p. 450.

17. Ibid., Preface and 1969, P I, p. 29 passim.

18. Esp. in 1969, throughout. The consequences of violating rules are stressed along with its inconsistency. Lewis's use of "pragmatic imperatives" and their relation to Kant are examined in detail in Saydah, R., *The Ethical Theory of Clarence Irving Lewis* (Athens: Ohio Univ. Press, 1969).

19. This is despite Lewis' tendency to treat animals in terms of behavior, not inner experience (1946, ch. 9, sect. 5–6, pp. 260–261). However, he recognizes that humans have elements of the animal in them (ibid.) and are also creatures of habit, like animals. By questioning the moral standing premise in Lewis, as Callicott does in Hume, subjective intrinsic value could be extended to at least some animals as "subjects of a life" as in Regan. The limiting of the scope of intrinsic value to humans could be questioned so as to include other species.

20. Ibid., ch. 17, sect. 7, p. 531. Lewis only touches upon this as a possible stance but seems to favor it.

21. See, e.g., K. Winetrout's book, *F. C. S. Schiller and the Dimensions of Pragmatism*, ch. 5 where he argues that pragmatism is "subjectivist" and thus neither idealist nor realist. This directly contradicts Peirce's characterization of his pragmaticism as "objective idealism," James's attempt to go beyond the subjective-objective distinction in experience, discussed later; Dewey's naturalism, which was an explicit critique of subjectivity, as I noted in chapter two; and Lewis's critique of emotivism in ethics, which argued for axiological realism, inter alia. Where I believe that Winetrout is correct is to note that Dewey moved in his transactional period closer to Schiller's humanism, in that the human element cannot be removed from knowledge. This point will be discussed in more detail.

22. "Some Consequences of Four Incapacities," reprinted in Charles S. Peirce, *Selected Writings*, ed. Wiener (New York: Dover, 1968), p. 69. As Wiener notes, "we must reject Descartes' method of universal doubt and appeal to self-evidence. Modern

science accepts no proposition as self-evident but rests on the consensus of the community of scientific investigators as to what premises one may adopt for the sake of inquiry" (ibid., p. 17). Peirce also argues that the community is an "end in itself," and thus connects community to value (ibid., p. 89). Cf. Callicott's notion of the "ecological community."

23. See Shaner, David E. and Duval, R. Shannon, "Conservation Ethics and the Japanese Intellectual Tradition," *Environmental Ethics*, 11/3, Fall 89, pp. 197–214. I am indebted to an essay by Rosenthal and Buchholz for bringing this article to my attention. See Rosenthal, Sandra B. and Buchholz, Rogene, A., "How Pragmatism Is an Environmental Ethic," in Light, A. and Katz, E., *Environmental Pragmatism* (New York: Routledge, 1996).

24. Ibid., Shaner and Duval. They also note the influence of Zen Buddhism on Nishida.

25. Ibid., p. 204.

26. Ibid., p. 212.

27. Ibid., p. 209.

28. Ibid. For a contrary view see Fuller, R., "American Pragmatism Reconsidered: William James Ecological Ethic," *Environmental Ethics*, 14/2, 1992.

29. Ibid.

30. Op. cit., Shaner and Duval. James began with a psychological approach to philosophy, understandable in a psychologist. However, his notion of experience attempts to go beyond the merely subjective. See, inter alia, "Does Consciousness Exist?" sect. II, p. 7 ff. and "The Idea of Consciousness" p. 120 in *Radical Empiricism*. (Reprinted in one volume with *A Pluralistic Universe*, New York: E.P. Dutton, 1971.)

31. Ibid., p. 274.

32. Norton has also labeled himself a pragmatist and defends "weak anthropocentrism." Whether this is to be interpreted as "subjectivist" or not is hard to say.

33. Peirce, "The Architecture of Theories," 1968, p. 153. Cf. p. 84 and 176–177.

34. I have previously documented this for Dewey and Lewis. The notion of community and consensus as regulative of inquiry in Peirce is holistic, since we do not start with universal doubt as a method, but have a body of agreed on beliefs.

35. Subjective intrinsic value theories exhibit a tension between individual incidents of value, whether in feeling or desire, and the bearer of value, the person or life as a whole. The incidents of value give life to the whole, but the whole is the condition of the incident of value. The relation to the whole is required, for neither feeling nor desire could take place except in a life of a subjective person. But the value of a life is measured by the feeling or desire. More, a lifetime is required for such subjective states in the highest degree or fullest extent. As this relation to a whole life is required, subjective theories are incomplete as theories of value if the relation is left out of the account. Judging the value of life as a whole requires such an account.

36. Indeed, analysis in terms of an object is a Cartesian residue, since an object opposed to a subject presumes the Cartesian metaphysics of a subject. This is rejected by the pragmatists.

37. Whether value is holistic or a part depends on context; sometimes value attaches to the whole, sometimes to a part of the whole.

38. As science is an activity, value and morals are not divorced from knowledge and the impact of scientific activity on the environment requires moral deliberation.

For the pragmatists value-knowledge of the environment is possible. Demonstration of actual value is somehow continuous with the rest of scientific investigations for the pragmatists.

39. See Dewey's letter to Bentley in John Dewey and Arthur F. Bentley, *A Philosophical Correspondence, 1932–1951*, ed., by S. Ratner and J. Altman (New Brunswick: Rutgers Univ. Press, 1964). Quoted in Rorty, R., *Consequences of Pragmatism* (Minneapolis: Univ. of Minnesota Press, 1982), ch. 5, p. 72.

40. Dewey and Bentley, *Knowing and the Known*, reprinted in *John Dewey, The Later Works: 1925–1953*, Vol. 16 (1949–1952), pp. 1–280. See esp. p. 260 and the rationale for the deliberate abandonment of the term 'experience' in n. 26, p. 73 at the end of ch. 2.

41. *Knowing and the Known* shares with Dewey's earlier work (1) contextualism, (2) a metaphysics of process, with events in time and space, (3) naturalism, (4) holism, (5) a rejection of both dualism and substantialism, as aspects of subjectivity, and (6) behaviorism. Only experience is superceded. I noted that Lavine argued that transactionalism undermined Dewey's earlier work. However, Dewey reinterpreted the categories of the earlier work in terms of transaction. Thus the connection of science and common sense is not undermined; both are seen as transactions (ibid., ch. 10, p. 242: written by Dewey alone). It is dualism that separates philosophy from human life (ibid., p. 249); transactionalism is the attempt to get them back together again.

42. Of course Peirce provides still another alternative, as I noted; Lewis, an alternative to subjective grounding (see n. 13).

43. This model was also challenged by W. James. See his essay, "Does Consciousness Exist?" in *Essays in Radical Empiricism*, ed. by R. B. Perry, 1943 (New York: E.P. Dutton, 1971). Bentley, however, saw that experience, not just consciousness, is problematic.

44. *Knowing and the Known*, ch. 4, p. 102 of reprint in op. cit., *John Dewey, The Later Works*. A knower detached from what is known, the subjectivist view, is replaced by transactional behavior as an event within process. See ibid., section 3, p. 100 ff.

45. See ibid., ch 5, p. 111 for the rejection of various sorts of dualisms.

46. See ibid., ch. 4, section 4. The attempt to treat reality in terms of process connects the project of *Knowing and the Known* with that of Whitehead in *Process and Reality*. There is some reference to the latter book in *Knowing and the Known*, but the influence of one on the other is unclear. There is also a resemblance to the flux of forces in Nietzsche's later metaphysics of the "will to power."

47. This may reflect the concurrent revolution in physics, relativity, in which velocity is relative to an observer (ibid. (sect. 4) p. 106 ff.). In the latter view there is no "absolute" velocity, since velocity requires a relation to an observer. The observer is thus in the relation, not outside of it. Niels Bohr concluded from this and also from quantum mechanics, that an epistemological revolution had been launched by contemporary physics.

48. See *Knowing and the Known*, ch. 10, p. 242 and 250 for Dewey's remarks on the human element in knowledge and the critique of idealist notions of objectivity. Of course Dewey might respond that he is overcoming the dualism of human and nature, i.e., that he is including the human observer within the transaction as a whole. I am indebted to K. Winetrout for drawing my attention to this connection in his fine book, *F.C.S. Schiller and the Dimensions of Pragmatism* (Athens: Ohio State Univ. Press, 1967).

49. Since our conceptual formulation is also in flux, transactionalism is historical. The concepts by which we formulate the flux themselves change. Human knowledge is finite and historical. This connects Dewey and Bentley to the historical forms of understanding in Heidegger and his followers. Although Rorty saw this connection, he did not realize that transactionalism made this connection. (For Rorty's views see the Preface.)

50. Loc. cit., *Knowing and the Known,* p. 260.

51. This essay is the first in the volume *Humanism, Philosophical Essays* (New York: Macmillan, 1903). The essay is reprinted in *Humanistic Pragmatism, The Philosophy of F.C.S. Schiller,* ed. Abel (New York: Free Press, 1966).

52. Although Schiller does not use the term "first" philosophy, ethics has that role in his view. His view is not simply a theory of values, but a fundamental axiology, or a first philosophy of the good. Ethics grounds metaphysics. This constitutes a novel view of first philosophy, in which it is distinct from metaphysics. I have argued this point in more detail in a forthcoming paper, "First Philosophy in the Pragmatic Humanism of F. C. S. Schiller."

53. Ibid., pp. 8–9. Cf. the essay "The Making of Reality," from *Studies in Humanism,* reprinted in 1966, P. I, ch. 5, p. 123. Schiller explored, like Peirce, the possibility of pan-psychism. See ibid., sect. 12, p. 143.

54. 1903, p. 8. Cf. p. 10 where value is the basis of fact.

55. Ibid., p. 4. Ethics is viewed as based on human good, i.e., a consequentialist model in which moral norms are derived from values, i.e., the human good. "Teleological valuation" is "the special sphere of ethical inquiry" thus pragmatism assigns validity to "the typical method of ethics" (ibid., p. 8).

56. I have attempted this project in my paper, "Toward a Deontological Environmental Ethics" in *Environmental Ethics,* 23/4, Winter 2001.

BIBLIOGRAPHY

Abel, R., ed., *Humanistic Pragmatism, The Philosophy of F. C. S. Schiller*, New York: Free Press, 1966.

Aiken, H., "Reflections on Dewey's Questions About Value," in Lepley, R., ed., *Value, a Cooperative Inquiry*, New York: Columbia Univ. Press, 1949, ch. 1.

Altman, A., "John Dewey and Contemporary Normative Ethics," *Metaphilosophy*, 13/2, 1982, 149–160.

Aristotle, *The Basic Works of Aristotle*, ed. McKeon, New York: Random House, 1941 (1970). Includes *Categories, Physics, Metaphysics, Nichomachean Ethics* and *Politics*. The translation of the *Nichomachean Ethics* is by W. D. Ross.

Attfield, R., *The Ethics of Environmental Concern*, New York: Columbia Univ. Press, 1983.

Ayer, A. J., *Language, Truth and Logic*, New York: Dover, 1952.

Ayres, C. E., "The Value Economy," in. Lepley, ed., *Value, A Cooperative Inquiry*, New York: Columbia Univ. Press, 1949, ch. 2.

Baylis, C., "Grading, Values and Choice," *Mind*, 67 (1958), p. 490.

Beardsley, M., "Intrinsic Value," *Philosophy and Phenomenological Research*, 26, 1965, p. 6 ff.

Bennett, J. O., "Beyond Good and Evil: A Critique of Richard Taylor's 'Moral Voluntarism,'" *Jour. of Value Inquiry*, 12/4, 1978, 313–319.

Bernstein, R. J., "Dewey's Naturalism," *Review of Metaphysics*, 13, December 1959, pp. 340–353.

———, *John Dewey*, New York: Washington Square Press, 1966.

Bourke, V. J., *History of Ethics*, Garden City: Doubleday/Image, 1968 (1970).

Callaway, H. G., "Democracy, Value Inquiry and Dewey's Metaphysics," *Jour. of Value Inquiry*, 27/1, 1993, pp. 13–26.

Callicott, J. Baird, "Animal Liberation: A Triangular Affair," *Env. Eth.*, 2/4, 1980, p. 316 (reprinted in *In Defense of the Land Ethic*, Albany: State University of New York Press, 1989, ch. 1).

———, "The Case Against Moral Pluralism," *Env. Eth.* 12/2, 1990.

———, "Elements of an Environmental Ethic: Moral Considerability and the Biotic Community" (reprinted in *In Defense of the Land Ethic*, ch. 4, p. 63).

———, *In Defense of the Land Ethic, Essays in Environmental Philosophy*, Albany: State University of New York Press, 1989.
———, "Intrinsic Value, Quantum Theory and Environmental Ethics," ch. 9 of *In Defense of the Land Ethics*, p. 157 ff.
———, "Non Anthropocentric Value Theory and Environmental Ethics," *Amer. Phil Quart.*, 21/4, 1984, p. 299.
———, "On the Intrinsic Value of Nonhuman Species," ch. 8 of *In Defense*, p. 131.
———, "Rolston on Intrinsic Value: a Deconstruction," *Env. Eth.*, 14/2, 1992.
Caspary, W. R., "Judgments of Value in John Dewey's Theory of Ethics," *Educ. Theory*, 40/2, 1990, 155–169.
Clement, R. C., "Watson's Reciprocity of Rights and Duties," *Env. Eth.*, 1/4, 1979.
Cooper, J., *Reason and Human Good in Aristotle*, Cambridge: Harvard Univ. Press, 1975.
Copleston, F., *A History of Philosophy*, Garden City: Doubleday/Image 1946 (1962); esp. Vol. 1, 5, 6 and 8.
Descartes, R., *Discourse on Method*, and *Meditations on the First Philosophy*, in one vol., *The Philosophical Works of Descartes*, trans. and ed. by Haldane and Ross, Cambridge: Cambridge Univ. Press, 1968.
Dewey, John, *Ethics* (coauthored with J. Tufts), New York: Henry Holt, 1908 (1932 rev.) [M5; L7].
———, *Reconstruction in Philosophy*, Boston: Beacon, 1920 (1972) [M12].
———, *Human Nature and Conduct*, New York: Modern Library, 1922 (1957) [M14].
———, "The Meaning of Value," *Jour. of Philosophy*, 22, Feb. 1925, pp. 126–133 (reprinted in *John Dewey, The Later Works*, ed. by Boydston, J., Carbondale: So. Illinois Univ. Press, 1984, pp. 98–103).
———, "Value, Objective Reference and Criticism," *Philos. Review*, 34, July 1925, pp. 313–332 (reprinted in *John Dewey, The Later Works*, ed. by Boydston, J., Carbondale: So. Illinois Univ. Press, 1984, pp. 78–97).
———, "The Ethics of Animal Experimentation," *Atlantic Monthly*, 138, Sept. 1926, pp. 343–346 (reprinted in *John Dewey, The Later Works*, ed. by Boydston, J., Carbondale: So. Illinois Univ. Press, 1984, pp. 98–103).
———, *Experience and Nature*, 1925, New York: Dover, 1929 ed. (1958) [L1].
———, *The Public and Its Problems*, Chicago: Swallow Press, 1927 (1953) [L2].
———, *The Quest for Certainty, A Study of the Relation of Knowledge and Action*, New York: Capricorn/ Putnam, 1929 (1960) [L4].
———, *Liberalism and Social Action* (reprinted in *John Dewey, The Later Works*, ed. by Boydston, J., Carbondale: So. Illinois Univ. Press, 1984, v. 11: 1–68).
———, *Freedom and Culture*, New York: Capricorn Books, 1939 [L13].
———, "Theory of Valuation," from the *International Encyclopedia of Unified Science*, Chicago: Univ. of Chicago Press, Vol. II, #4, 1939 [L13].
———, "Experience, Knowledge and Value, A Rejoinder," in Schilpp, P. A., ed. *The Philosophy of John Dewey*, New York: Tudor Publ., 1939 (1951), pp. 517–608.
———, "Ambiguity of Intrinsic Good," *Jour. of Philosophy*, 39, June 1942, pp. 328–330 (reprinted in *John Dewey, The Later Works, Vol. 15, 1942–1948*, ed. by Boydston, J., Carbondale: So. Ill. Univ. Press, 1989, p 42 ff.).
———, "Valuation Judgments and Immediate Quality," *Jour. of Philosophy*, 40, June 1943, 309–317 (reprinted in *John Dewey, The Later Works, Vol. 15, 1942–1948*, ed. by Boydston, J., Carbondale: So. Ill. Univ. Press, 1989, p 63 ff.).

―――, "Further as to Valuation as Judgment," *Jour. of Philosophy*, 40, Sept. 1943, 309–317 (reprinted in *John Dewey, The Later Works, Vol. 15, 1942–1948*, ed. by Boydston, J., Carbondale: So. Ill. Univ. Press, 1989, p 73 ff.).

―――, "Some Questions about Value," *Jour. of Philosophy*, 41, August 1944, pp. 449–455 (reprinted in *John Dewey, The Later Works, Vol. 15, 1942–1948*, ed. by Boydston, J., Carbondale: So. Ill. Univ. Press, 1989, p 101 ff.).

―――, "The Field of Value," in Lepley, R., ed., *Value, a Cooperative Inquiry*, New York: Columbia Univ. Press, 1949, ch. 3, p. 66 (reprinted in *John Dewey, The Later Works, Vol. 16, 1949–1952*, ed. by Boydston, J., Carbondale: So. Ill. Univ. Press, 1989, p. 343 ff.).

―――, with Arthur Bentley, *Knowing and the Known*, reprinted in *John Dewey, The Later Works, Vol. 16, 1949–1952*, ed. by Boydston, J., Carbondale: So. Ill. Univ. Press, 1989, pp. 1–280.

Edel, A., "The Concept of Value and its Travels in Twentieth Century America," in Murphey, M. G. and Berg, I., eds., *Values and Value Theory in Twentieth Century America*, Philadelphia: Temple Univ. Press, 1988.

Ehrlich, P. R. and A. H., *Betrayal of Science of Reason, How Anti-Environmental Rhetoric Threatens our Future*, Washington: Island Press, 1998.

Ferm, V., ed., *A History of Philosophical Systems*, New York: Philosophical Library, 1950.

Findlay, J. N., *Axiological Ethics*, London: Macmillan, 1970.

Frankena, W., *Ethics*, Englewood Cliffs: Prentice-Hall, 1963.

Frey, R. G., "Why Animals Lack Beliefs and Desires," in *op. cit., Animal Rights and Human Obligations*, P II, sect. 11.

Frondisi, R., *What Is Value—An Introduction to Axiology*, trans. Lipp, La Salle: Open Court Publ., 1971.

Fuller, R., "American Pragmatism Reconsidered: William James Ecological Ethic," *Env. Eth.*, 14/2, 1992.

Garnett, A. C., "Intrinsic Good : its Definition and Referent," in Lepley, R., ed., *Value, a Cooperative Inquiry*, New York: Columbia Univ. Press, 1949.

Garrison, J., "John Dewey, Jacques Derrida and the Metaphysics of Presence," *Transactions of the Charles S. Peirce Society*, 35/2, 1999, 346–372.

Godfrey-Smith, W., "The Value of Wilderness," *Env. Eth.*, I/4, 1979, 309.

Goodpaster, K., "On Being Morally Considerable," *Jour. of Phil.* 75, 1978.

Goodpaster, K. E. and Sayre, K. M., eds., *Ethics and the Problems of the 21st Century*, Notre Dame: University of Notre Dame Press, 1979.

Gottlieb, R. S., ed., *The Ecological Community*, New York: Routledge, 1997.

Gouinlock, J., *John Dewey's Philosophy of Value*, New York: Humanities Press, 1972.

―――, ed., *The Moral Writings of John Dewey*, New York: Macmillan/Hafner, 1976.

―――, "Dewey," in *Ethics in the History of Western Philosophy*, ed. Cavalier, R., Gouinlock, J., and Sterba, H. P., New York: St. Martin's Press, 1989, ch. 10, pp. 306–332.

―――, "Dewey's Theory of Moral Deliberation," *Ethics*, 88/3, 1978, 218–228.

Greenstein, H., "Intrinsic Values and the Explanation of Behavior," *Jour. of Value Inquiry*, 4/4, 1971, 304.

Gunn, A., "Why Should We Care About Rare Species," *Env. Eth.*, II/1, 1980.

Hartshorne, C., "The Rights of the Subhuman World," *Env. Eth.*, I/1, 1979, 49.

Hoff, C., "Kant's Invidious Humanism," *Env. Eth.*, 5/1, 1984, 63–70.

Holmes, R. L., "John Dewey's Moral Philosophy in Contemporary Perspective," *Rev. Meta.*, 30/1, (77), 1966, 42–70.

Honneth, A., "Between Proceduralism and Teleology: An Unresolved Conflict in Dewey's Moral Theory," *Transactions of the Charles S. Peirce Society*, 34/3, 1998, 689–711.

Hook, S., "The Ethical Theory of John Dewey," in *The Quest for Being and Other Studies in Naturalism and Humanism*, New York: St. Martin's Press, 1961, pp. 49–70.

Hume, D., *Treatise of Human Nature*, reprinted in *Hume's Moral and Political Philosophy*, ed. Aiken, Darien: Hafner, 1970.

——— , *An Enquiry Concerning the Principles of Morals*, reprinted in *Hume's Moral and Political Philosophy*, ed. Aiken, Darien: Hafner, 1970.

Hurley, P. E., "Dewey on Desires: The Lost Argument," *Transactions of the Charles S. Peirce Society*, 24/3, 1988, 509–519.

James, W., *Radical Empiricism*. (Reprinted in one volume with) *A Pluralistic Universe*, New York, E. P. Dutton, 1971.

——— , *Pragmatism and Other Essays*, 1910, New York: Washington Square Press, (1975)

Jessup "On Value," in Lepley, R., ed., *Value, a Cooperative Inquiry*, New York: Columbia Univ. Press, 1949.

Jonas, Hans, *The Imperative of Responsibility, In Search of an Ethics for the Technological Age*, Chicago: Univ. of Chicago Press, 1984.

Kant, I, *Groundwork of the Metaphysics of Morals*, Abbott trans., London: Longman, Green and Co., 1785 (1873).

——— , *Critique of Practical Reason*, trans. Beck, L., Indianapolis: Bobbs-Merrill, 1788 (1956, 1976 ed.).

——— , *Metaphysics of Morals*, trans. Ladd, Indianapolis: Bobbs-Merrill, 1797a (1965).

——— , The "Doctrine of Virtue," the second part of the *Metaphysics of Morals*, trans. Ellington, J., Indianapolis, Hackett, 1797b (1994).

——— , *Lectures on Ethics*, trans. Infield, L., Indianapolis: Hackett, 1775–80 (1930, 1963 ed.).

Katz, E., "Imperialism and Environmentalism," in Gottlieb, R.S., ed., *The Ecological Community*, New York: Routledge, 1997, pp. 163–174 (ch. 8).

——— , "Searching for Intrinsic Value: Pragmatism and Despair in Environmental Ethics," reprinted in *Environmental Pragmatism*, 1996, ch. 15, p. 307 ff. Reprinted from *Env. Eth.*, 9/3, 1987.

Katz, E., and Light, A. eds., *Environmental Pragmatism*, New York: Routledge, 1996.

Khatchadourian, H., "'Intrinsic' and 'Instrumental' Good: an Untenable Dichotomy," *Jour. of Value Inquiry*, 4/3, 1969, 172.

Kheel, Marti "Nature and Feminist Sensitivity," in *Animal Rights and Human Obligations*, Englewood Cliffs: Prentice-Hall, 1976, p. 256 ff.

Kockelmans, J., ed., *Contemporary European Ethics*, Garden City, Doubleday/Anchor, 1972.

Lee, D. S., "Pragmatism and Natural Values," *Journal of Value Inquiry*, 17, 1983, 191–202.

Lee, H. N., "Methodology of Value Theory," in Lepley, R., ed., *Value, a Cooperative Inquiry*, New York: Columbia Univ. Press, 1949, ch. 8.

Lemos, N. M., *Intrinsic Value, Concept and Warrant*, Cambridge: Cambridge Univ. Press, 1994.

Lepley, R., ed., *Value, a Cooperative Inquiry*, New York: Columbia Univ. Press, 1949.
Lewis, C. I., *An Analysis of Knowledge and Valuation* (The Paul Carus Lectures Series 7), La Salle, Ill.: Open Court, 1946 (1962 ed.).
——, *Values and Imperatives*, ed. Lange, J., Stanford: Stanford University Press, 1969.
Lewis, H., *A Question of Values: Six Ways We Make the Personal Choices that Shape Our Lives*, San Francisco: Harper, 1990
Mackay, D. S., "Pragmatism," ch. 31 in Ferm, V., ed., *A History of Philosophical Systems*, New York: Philosophical Library, 1950, pp. 381–404.
Mead, G. H., *Selected Writings*, ed. Reck, A. J., Indianapolis: Bobbs-Merrill, 1964.
Merchant, C., "Environmental Ethics and Political Conflict: A View from California," *Env. Eth.*, 12/1, 1990.
Mill, J. S., *Utilitarianism*, in *The Utilitarians*, Garden City: Anchor/Doubleday, 1863 (1973), in an edition with *On Liberty*, and Bentham's *Principles of Morals and Legislation*.
Miller, P., "Value as Richness: Toward a Value Theory for an Expanded Naturalism in Environmental Ethics," *Env. Eth.*, 4/2, 1982.
Mitchell, E. T., "Dewey's Theory of Valuation," *Ethics*, July 1945, 55, 287–297.
——, "Values, Valuing and Evaluation," in Lepley, ed. *Value, a Cooperative Inquiry*, New York: Columbia Univ. Press 1949, p. 202 ff.
Moore, G. E., *Principia Ethica*, Cambridge: Cambridge Univ. Press, 1903 (1959 ed.).
——, *Ethics*, Oxford: Oxford Univ. Press, 1912 (1965).
——, "Is Goodness a Quality?" from *Philosophical Papers*, New York: Macmillan, (1959) 1963.
——, "The Conception of Intrinsic Value" in *Philosophical Studies*, London: Routledge, 1922 (1958).
——, "Nature of Moral Philosophy," ch. 10 of *Philosophical Studies*, London: Routledge, 1922 (1958).
Nash, R. F., *The Rights of Nature*, Madison, University of Wisconsin Press, 1989.
Neville, R., *The Highroad Around Modernism*, Albany: State University of New York Press, 1992.
Nielsen, Kai, "On Ascertaining What is Intrinsically Good," *Jour. of Value Inquiry*, 10/2, 1976.
Nietzsche, F., *Beyond Good and Evil*, trans. Kaufman, New York: Vintage, 1966.
——, *Will to Power*, trans. Kaufmann, New York: Vintage/Random House, 1967.
Nissen, B., "John Dewey on Means and Ends," *Phil. Research Archives*, 3, 1977, 709–738.
Norton, B., "The Constancy of Leopold's Land Ethic," reprinted in Katz, E., and Light, A. eds., *Environmental Pragmatism*, New York: Routledge, 1996.
——, "Integration or Reduction, Two Approaches to Environmental Values," in Katz and Light *Environmental Pragmatism* 1996, ch. 5.
——, *Why Preserve Natural Variety?*, Princeton: Princeton Univ. Press, 1987.
Osborne, H., *Foundations of the Philosophy of Value*, Cambridge: Cambridge Univ. Press, 1933.
Pappas, G. F., "To Be or To Do, John Dewey and the Great Divide in Ethics," *Hist. Phil. Quart.*, 14/4, 1997, 447–472.
Parker, D. H., "Discussion of John Dewey's 'Some Questions About Value,'" in Lepley, R., ed., *Value, a Cooperative Inquiry*, New York: Columbia Univ. Press, 1949.

Parker, K., "Pragmatism and Environmental Thought," Katz and Light, *Environmental Pragmatism*, New York: Routledge, 1996, ch. 1.
——, "The Values of a Habitat," *Env. Eth.*, 12/4, 1990.
Charles S. Peirce, Selected Writings, ed. Wiener, P., New York, Dover, 1968.
Pepper, S., "Some Questions on Dewey's Esthetics," in Schilpp, P.A., ed. *The Philosophy of John Dewey*, New York: Tudor Publ., 1939 (1951), pp. 371–389 (sect. 12).
——, "Observations on Value from an Analysis of Simple Appetition," in Lepley, R., ed., *Value, a Cooperative Inquiry*, New York: Columbia Univ. Press, 1949. ch. 13.
——, "A Brief History of General Theory of Value," in Ferm, V., ed., *A History of Philosophical Systems*, New York: Philosophical Library, 1950.
Perry, R. B., *Realms of Value, A Critique of Human Civilization*, Cambridge: Harvard University Press, 1954.
Phihar, E. B., "The Justification of an Environmental Ethic," *Env. Eth.*, 5/1, pp. 47–61.
Plato, *Collected Dialogues*, Cairns, H. and Hamilton, E., eds., Princeton: Princeton University Press, 1973 ed. Including *Apology, Crito, Gorgias, Meno, Phaedo, Philebus, Protagoras, Republic*, and *Symposium*.
Potter, V. J., *Charles S. Pierce on Norms and Ideals*, Worcester: University of Massachusetts Press, 1967.
Povilitis, A. J., "On Assigning Rights to Animals and Nature," *Env. Eth.*, 2/1, 1980, 67–71.
Prall, David, W., *A Study in the Theory of Value*, Berkeley: Univ. of Calif. Publications in Philosophy, 3/2, 1921, pp. 179–290.
Rachels, J., "Darwin, Species and Morality," in *Animal Rights and Human Obligations*, Regan, T., Singer, P., eds., part 3, sect. 3.
Reagan, T., "Animal Rights and Human Wrongs," *Env. Eth.*, II/2, 1980, 99 ff.
——, "The Case for Animal Rights," in *Animal Rights and Human Obligations*, ed. T. Regan and P. Singer, Englewood Cliffs: Prentice Hall, 1976b.
——, "The Nature and Possibility of an Environmental Ethic," *Env. Eth.*, 3/2, 1981, 19–34.
——, "Why Death Does Harm to Animals," in *Animal Rights and Human Obligations*, Regan, T., Singer, P., eds., Englewood Cliffs: Prentice-Hall, 1976c, p. 153 ff.
Regan, T., and Singer, P., eds., *Animal Rights and Human Obligations*, Englewood Cliffs: Prentice-Hall, 1976a.
Rescher, N., *Introduction to Value Theory*, Englewood Cliffs: Prentice-Hall, 1969.
Rice, P., "Science, Humanism and the Good," in Lepley, ed., *Value, a Cooperative Inquiry*, New York: Columbia Univ. Press, 1949, ch. 14.
Richmond, S. A., "On Replacing the Notion of Intrinsic Value," *Jour. of Value Inquiry*, 10/3, 1976, 205.
Rolston, Holmes III, "Are Values in Nature Subjective or Objective?" *Env. Eth.* 4/2, 1982, 134.
——, *Environmental Ethics, Duties to and Values in the Natural World*, Philadelphia: Temple Univ. Press, 1988.
——, "Feeding People versus Saving Nature," in Gottlieb, R. S., ed., *The Ecological Community*, New York: Routledge, 1997, p. 208.

Rosenthal, Sandra B. and Buchholz, Rogene, A., "How Pragmatism Is an Environmental Ethic," in Light, A. and Katz, E., eds., *Environmental Pragmatism*, New York: Routledge, 1996.
Russow, L.-M., "Why Do Species Matter?" *Env. Eth.*, 3/2, 101–112.
Sahakian, W. S., *Ethics, An Introduction to Theories and Problems*, New York: Harper and Row/Barnes and Noble, 1974.
Saydah, R., *The Ethical Theory of Clarence Irving Lewis*, Athens: Ohio Univ. Press, 1969.
Scheler, M., *Formalism in Ethics and a Non-formal Ethics of Values*, trans. Frings and Funk, Evanston: Northwestern Univ. Press, 1916 (1973).
Schiller, F. C. S., *Humanism, Philosophical Essays*, New York: Macmillan, 1903.
———, *Humanistic Pragmatism, The Philosophy of F. C. S. Schiller*, ed. Abel, New York: Free Press, 1966 which includes a reprint of the essay, "The Making of Reality," from *Studies in Humanism* (ch. 5).
Schilpp, P., ed, *The Philosophy of John Dewey*, New York: Tudor Publ., 1951.
Schweitzer, A., *Civilization and Ethics*, trans. J. Naish, reprinted in *Animal Rights and Human Obligations*, ed. T. Regan and P. Singer, Englewood Cliffs: Prentice-Hall, 1976, sect. 10.
Shaner, David E. and Duval, R. Shannon, "Conservation Ethics and the Japanese Intellectual Tradition," *Env. Eth.*, 11/3, 1989, 197–214.
Singer, P., *Animal Liberation*, New York: New York Review Press, 1975.
Smith, John E., *The Spirit of American Philosophy*, London and New York: Oxford University Press, 1963.
Stuart, H. W., "Dewey's Ethical Theory," in Schilpp, P. A., ed. *The Philosophy of John Dewey*, New York: Tudor Publ., 1939 (1951), pp. 293–333 (sect. 10).
Taylor, A. E., *Plato: The Man and His Work*, London: Methuen, 1926 (1978).
Taylor, Bob P., "John Dewey and Environmental Thought: A Response to Chaloupka," *Env. Eth.*, 12/2, 1990, 175.
Taylor, Paul W., "The Ethics of Respect for Nature," *Env. Eth.*, 3/3, 1981.
———, "In Defense of Biocentrism," *Env. Eth.*, 5/3, 1983, 237–243.
Taylor, R., *Action and Purpose*, Englewood Cliffs: Prentice-Hall, 1966.
Tiles, J. E., *Dewey*, London/New York: Routledge, 1988.
Vallentyne, P., "The Teleological-Deontological Distinction," *Jour. of Value Inquiry*, 21/1, 1987.
Vivas, E., *The Moral Life and the Ethical Life*, Chicago: Univ. of Chicago Press, 1950, esp. ch. 6 and 7, "The Instrumental Moral Theory," I and II.
Wellbank, J. H., "Is a New Definition of Ethical Naturalism Needed?" *Jour. of Value Inquiry*, 8/1, 1974, 46 ff.
Werkmeister, W., *Historical Spectrum of Value Theories*, Lincoln: Johnsen Publ. Co., 1970.
Weston, "Before Environmental Ethics," Light, A. and Katz, E., eds., *Environmental Pragmatism*, New York: Routledge, 1996, ch. 7, p. 139 (reprinted from *Env. Eth.*, 14/4, 1992).
———, "Beyond Intrinsic Value, Pragmatism in Environmental Ethics," Light, A. and Katz, E., eds., *Environmental Pragmatism*, New York: Routledge, 1996, ch. 14, p. 285 (reprinted from *Env. Eth.*, 7/4, 1985).
White, Morton, "Value and Obligation in Dewey and Lewis," *Philosophical Review*, 58, July 1949, 322.

Willard, L. D., "Intrinsic Value in Dewey," *Phil. Research Archives*, I, 1975, 54–77.
——, "On Preserving Nature's Aesthetic Features," *Env. Eth.*, 2/4, 1980.
Windelband, W., "History and Natural Science" (Rectorial Address), trans. Oakes, in *History and Theory*, 1980, xix.
——, *A History of Philosophy*, trans. Tufts, New York: Harper and Bros., 1891 (1901, 1958 ed.).
Zimmerman M., "The Critique of Natural Rights and the Search for a Non-Anthropocentric Basis for Moral Behavior," *Jour. of Value Inquiry*, 19/1, 1985.

INDEX

actions, 2, 22, 33, 38, 82, 103, 111, 113, 122, 125, 127–28, 138
activities, 37, 68, 75, 80–82, 84–87, 94–95, 97, 100, 104–6, 109–10, 112, 116, 119, 125, 127, 129–31, 138–39, 141, 155
aesthetic. *See* esthetic
Agassiz, Louis, 149
air, soil and water, 15, 18, 25–28, 35, 40, 72, 126, 128–30, 132, 138
altruism, altruistic, 22, 29–30, 35
analysis, vi, vii, viii, 18, 20–21, 32, 50, 52, 54, 63, 65, 72, 91, 100–1, 105–7, 110, 121, 140, 150, 154
animals, v–vii, 1–5, 7–13, 15–18, 20, 22, 24–25, 27–28, 30, 32–33, 35–36, 43–48, 50, 53, 55–56, 58, 73, 81, 85–88, 124, 129–30, 132, 134, 143, 148, 153
   a. rights, 2, 3
   theories of, 2
anthropocentric, v–viii, 2–4, 9, 11–14, 16, 22, 29, 34, 37–38, 45, 49–53, 56, 57–59, 61–62, 67, 72, 79, 81–83, 88–89, 121, 131, 134, 143–44, 147–48, 150, 156
   a. ethics, 3, 9, 12–13, 29, 34, 37
   a. value theories, v
appraisal, 47, 98, 100, 111, 113, 131
arbitrary, v, 20, 77, 97, 99, 117, 127, 141

arguments
   extension, vii, ix, 7–8, 13–15, 17, 22, 24–31, 35, 45, 56, 79, 88, 105, 123, 133–34, 139, 144, 148, 156
   from marginal cases, 7, 8, 53
Aristotle, Aristotelian, 20, 44, 46, 50, 91–93, 96, 100, 102, 105–6, 116, 120, 154, 157
art, 42, 68, 121, 136
atomism, 151
attitude(s), 60
attributes, 1, 3, 91
autonomy, vi, ix, 50, 157
axiology, axiological, 8–9, 15–16, 19, 34, 36, 38, 43, 50–51, 156
   a. ethic, 8, 36, 38

basis, vi, 1, 3–5, 7–8, 13, 16, 20, 22–23, 25, 29–31, 34–35, 38, 50–51, 53, 56–58, 61–62, 68, 75, 79–80, 83, 86, 88, 96, 109, 117, 119, 123, 125, 130–31, 133, 140, 143–49, 151, 155–57
bearer, v, 3, 5–6, 8, 10–11, 23, 27–28, 39, 44, 50–51, 81–85, 88, 96, 98, 111, 120, 130, 146, 156
   of rights, v
   of values, 3, 55
beauty, beautiful, 29, 39–40, 44, 51, 76, 136–37, 145, 147

INDEX

behavior, behaviorism, 22, 30–31, 72–73, 77, 80–81, 83–88, 92–93, 96, 104, 122, 132–33, 153–55
being, vi, 4, 6–7, 10, 12, 15–16, 21–24, 26, 31, 43, 48, 52–53, 58, 60, 63, 69, 71–73, 75, 78, 82–84, 88, 91, 93–95, 98, 100, 107, 109, 120, 126, 137, 148, 155
belief, 52, 60, 78, 82, 102
benefit-cost analysis, (BCA) 52
Bentham, Jeremy, 8–9
Bentley, Arthur, viii–x, 153–55
biocentric theories, 2
Biodiversity, 133
biology, biological, 13, 25, 29–30, 37, 72, 74, 79, 82–84, 86, 128–29, 139
 b. nature, 72
biosphere, v–vi, 1–2, 15–17, 20, 27, 30, 32, 38, 50, 58, 85, 88, 128–29, 132–33, 138, 152
biotic community, -ies, 11, 14–16, 20, 22–25, 27–31, 35–37
body, 72–75, 81–82, 86, 94
Bowers, C. A., 57, 59–61, 89
bringing about (ends/values).* *See* value, Dewey's theory of, as "bringing about"

Callicott, J. Baird, vi, viii, 1, 7, 9–42, 45–51, 53–56, 58–59, 65, 121, 123–24, 134–35, 140, 146, 148, 152
capacity, v, 2, 26, 40, 54, 136
care, 22, 25, 114, 135, 137
caring for, 83, 86–87, 95, 125–26, 129, 131–32, 139
Cartesian, 12, 14–15, 17, 19, 37–38, 49, 54, 63–65, 68, 71, 73–74, 79, 82, 89, 108, 120, 124, 144, 148–51, 153. *See also* Descartes
cause(s), 69–70, 76, 83, 93, 104, 107, 110, 112, 114, 116, 143, 146
 c. and effect, 69–70, 76, 104, 107
change(s), vi, ix, xi, 9, 15–16, 19, 23, 47, 68–69, 71, 76–78, 84, 104, 108, 111, 120, 123, 125–28, 141, 153
character, 2, 17, 21–23, 26, 34, 47, 55, 69–72, 74–75, 81, 87, 99, 104, 106, 108, 110–11, 113, 115, 117–20, 129, 146
 c. of nature, 69
choice, 48, 94, 97, 113, 116, 136
circumstances, 35, 44, 47, 64, 78, 95, 97–100, 102, 106, 111–13, 115–16, 119, 121–22, 153
cognitivism, 145
coherentist, ix–x
communication, 79, 87, 133
 social, 72
community, communities, viii, 14, 16–17, 25, 27–29, 31–32, 36–37, 39, 50, 138–39, 145, 148–50
 biotic, 11, 14–16, 20, 22–25, 27–31, 35–37
 moral, 29, 32
conativism, 27
concepts, 50, 82
conclusion, 8, 10, 24, 94
condition(s), 3–4, 28–29, 31–32, 37, 42–44, 48, 50, 53, 61, 70, 72, 78–80, 84, 87, 92–94, 99, 103, 106, 109, 111–14, 117–18, 123, 125, 127, 129–30, 132, 134–35, 147, 153
 social, 111, 121
conduct, 77–78, 81, 84–85, 117–18
connections, 70–71, 83–85, 92, 94–97, 111, 119–20, 130
conscious, 4–5, 12, 18–19, 33, 41, 49, 57, 71–72, 123, 150
consciousness, 3–4, 13, 17–21, 23, 25, 27, 41, 43, 51, 54, 60, 82, 130, 149–50
 stream of, 73, 149–50, 154
consequence(s), 5, 7, 30, 34, 62, 70, 84–86, 93–94, 98–101, 103–4, 115–18, 124, 128, 132, 148, 153
 environmental, 128, 140
 personal, 138
 social, 85, 116, 118, 128, 138, 140
Consequences of Pragmatism, x
consequentialism, viii–ix, 104–5, 116–19, 121, 145
consequentialist, vii, 35, 48, 56, 100, 104, 115–16, 119, 124–25, 147, 152
 c. ethic. *See* ethic, consequentialist
consistency, 20, 33, 148

Index

consolation, xi
consummation, 70, 76–77, 95–96, 98
   c. of experience, 76
contemplation, contemplative, xi, 65, 106, 127, 156
content, 55, 113, 131
context, contextualism, x, 6, 9, 12, 31, 47, 49, 54, 59, 62, 64, 67, 79–80, 91–92, 99, 111, 113–14, 116, 129, 141, 144, 147, 150, 152, 156
contract, 3, 8, 23
convention, conventional, 21, 34, 58
creation, 58, 65, 70, 83, 135
critical, 3, 20, 31, 91, 95–96, 98, 102, 105, 109, 113, 123, 143
culture(s), cultural, 15, 37–39, 42–45, 47, 49, 57, 60–61, 64, 71, 74, 79
custom, 71, 81
cycle(s), 151

Darwin, Charles, 22, 29, 30, 55, 58, 88
data, 105
decision(s), 60, 113–14, 134, 136
deep ecologists. *See* ecologists, deep
definition, 2, 18, 103, 112, 115, 129, 155
degrees of value. *See* value, degrees of
deliberation, 52, 76, 81, 93–94, 96, 98–99, 104–6, 112–13, 115, 119, 121, 125–26, 128, 135, 140
   moral, 115, 117, 119, 121, 125, 139
demand value, 51
democracy, 61
deontological, 48
Descartes, Rene, 13, 15, 19, 23, 48, 151–52. *See also* Cartesian
description, descriptive, 31, 41, 93, 154
desirable, 63, 94, 98, 103, 112–13, 136
desire(s), desiring, 20, 23, 26, 32, 34, 41, 63–65, 73, 81–82, 94, 99–100, 103, 105, 107, 110–12, 114–16, 131, 138, 144
   human, 47, 61, 88, 131
determinate, 93, 98, 116
devaluation, viii, 12, 55, 89, 95, 97, 99, 144, 155–56
   of nature, viii, 89, 107–8, 155–56
   of the natural world, 55

development, 22, 29, 32, 52, 59, 68, 87, 89, 101, 116, 118, 120, 137–38, 150
   economic, 137–38
Dewey, John, vi–xi, 36–37, 42, 44, 49–50, 56
   and pragmatism, 143–47, 150–57
   his critics in the environmental movement, 57–65
   his ethics used as a basis for deciding environmental issues, 123–41
   holism, 109–22
   naturalism, 67–89
   theory of value and instrumentality, 91–108
   See also values, theory of, Dewey
difference, vii–viii, xi, 2, 7, 14–15, 22, 30, 49, 51, 68, 71, 87, 109, 136
dimension, 11, 24, 43
distinction, 2, 11, 18–20, 23, 49, 52, 54, 63, 65, 68, 71–72, 87, 92–94, 102, 105–6, 112, 121, 144, 147–48, 150–52, 154
   fact-value, 48–49, 54
   is and ought, 31
   subjective-objective, 19
duty, v, 8–9, 15, 35–38, 49–50, 124, 129
Duval, R. Shannon, 149

Earth, 37, 43
ecocentric theories, 2
ecocentric world view, 149
ecologists, 52
   deep, viii, 26, 152
ecology, vii, 2, 14–16, 27, 29–30, 32–33, 35, 39, 48, 53, 59, 149, 151
   science of, 14, 23, 29–30
economic(s), 28, 41–44, 55, 126, 131, 133, 136–38
economic development. *See* development, economic
ecosystems, v–vi, 1–2, 9, 11, 16, 24, 36–37, 39–40, 42, 48, 50, 53–54, 56
effect, 9, 26, 30, 32–33, 37, 50, 59, 64, 68–71, 78–79, 93, 104, 106, 111, 116, 125, 135, 146. *See also* cause and effect

ego, egoism, 13, 21–22, 25, 149
egocentric, egocentrism, 25, 149
element(s), 65, 75, 79, 81–83, 88–89, 92, 110–16, 118–19, 125, 129, 132, 135, 137, 149, 152, 154
emotion(s), 20, 91
empiricism, 122, 149, 153
empiricist(s), 83
end(s), xi, 7–9, 11, 29, 40, 46, 47, 51, 53, 59, 64–65, 68–70, 74–77, 82, 85–86, 88, 92–108, 111–12, 114–19, 122, 127, 131–33, 135–38, 140, 145, 147–48
    fixed, 76, 89, 97, 102–3, 105, 118, 120–21, 131
    isolated, 97
    overall, 94, 96, 127
    transcendent, 96, 102
end in itself, 9, 40, 47, 53, 93, 95, 98, 103, 106
end(s)-in-view, 92–94, 96, 99, 100, 102, 105–7, 111–16, 118–19, 121, 125, 127, 132
endangered species. *See* species, endangered
energy, 14, 135
entity, 19–24, 34, 55, 61, 157
environment, v–vi, viii, 2–3, 5, 8, 12–14, 16, 20, 24–26, 28–29, 37–40, 43, 47–48, 50, 52, 54, 55, 57–63, 68–69, 72–86, 88–89, 98, 101, 108, 116, 118, 120–22, 123–35, 137–41, 143–44, 147–57
    as a whole, 26, 28, 78, 126–27, 130–32, 135, 138–39, 143, 156
    exploitation of the e., 13
    impact on the, of science, et al., 38, 47, 59, 118, 120, 124, 126–29, 131–32, 135, 140–41, 152, 156–57
    physical, 78, 80, 128, 139–40
    social, 74–75, 79–80, 87, 107, 109, 128, 132, 138–40, 150
    unique, 137
environmental ethic(s). *See* ethics, environmental
environmental consequences. *See* consequences, environmental
environmental philosophers. *See* philosophers, environmental
environmentalism, vi
Epicurus, x
epistemological, 34, 55, 60
epistemology, 9, 89
equality, 6, 10, 31, 36, 54, 133–34
essence, 2, 34, 75, 91
esthetic(s), 9, 11, 69, 75, 97, 102, 137, 144–45, 147–49, 152
    esthetic theories of intrinsic value, 9
ethic(s), ethical, v–viii, x 1, 3–4, 8–9, 11–17, 20, 22, 24–33, 35–39, 46–48, 50–51, 54–56, 57, 61–62, 63–65, 67–68, 76, 79, 89, 103–5, 119, 121, 123–24, 129–30, 140–41, 143–44, 148–49, 151–52, 155, 156
    anthropocentric, viii, 3, 9, 12–14, 29, 34, 37
    axiological, 8, 9, 36, 38, 156
    character, e. of perfection of, 120
    consequentialist, vii, 124. *See also* consequentilism
    environmental, v–ix, 1–3, 5, 8–9, 12–17, 24, 26, 33, 37–38, 50, 57–58, 61–64, 67–68, 79, 80, 91, 123–25, 128–29, 134, 139–40, 143–44, 148, 152–57
    management, 3–4
    Western, vi, 12, 46, 102
"Ethical Basis of Metaphysics, The" (essay), 156
Ethics (Dewey and Tufts), ix
evaluation(s), 41, 52, 76, 81, 94–95, 98–100, 104–6, 112–13, 115–16, 118, 121, 126–27
evaluative, 19, 40
event, x, 38, 67, 69–70, 72–76, 80, 83, 96, 108, 122, 154, 155
evidence, ix–x, 6, 24, 29, 35, 38, 49, 69, 72, 74, 86, 113, 145, 148–49
evil, 69, 75, 99
evolution, 22, 35, 43–44, 46, 70, 72, 74, 77, 88
    theory of, 29

existence, 21, 30, 52, 55, 68–70, 72, 75–77, 83, 97, 100, 102, 105–7, 114, 120, 149
experience, x–xi, 42, 51–52, 59–60, 62–63, 65, 67, 70–77, 79, 81, 83, 89, 95–96, 100, 103–4, 106, 110–11, 113, 125, 133, 137, 145–56
    consummation of, 76–77, 95, 98
experiment(s), 112
experimentalism, experimentalist, 102, 126
experimental method, 113, 115
extension argument. *See* argument, extension
extrinsic, 2, 38, 77, 92, 96, 99, 103, 145–47

fact(s), viii, 18–20, 23, 31, 39, 43, 47, 49, 54, 64, 72, 80, 97, 99, 107, 111, 112, 132, 155
    fact-value distinction. *See* distinction, fact-value
faculty, faculties, 2, 19, 82, 88
feeling(s), 4, 20, 23, 31–32, 34, 38, 79, 81–82, 91, 104, 151
field, v, 8, 13, 73, 83, 107, 157
fixed ends. *See* ends, fixed
flux, 78, 153, 154
folkways, 61, 109
food chain, 14, 17, 24, 27–28, 30, 39, 45, 137
force(s), 62, 81, 88
for its own sake, for their own sake, 2, 11, 19, 21, 42, 103, 146–47
for the sake of, 2, 11, 32, 103–4, 111, 132
form, v, xi, 1, 5, 8, 10, 13, 20, 34, 39, 40, 42–46, 48, 51, 54, 57, 60, 62, 68–72, 77, 79–80, 82, 87, 89, 91, 94, 99, 111, 114–15, 117–19, 121, 125, 133–35, 144, 154–55, 157
fossil fuels, 135
foundational(ism), vi–viii, 9, 14, 30, 38, 76–78, 82, 89, 105, 114, 119–20, 123–24, 127, 134
    grounding, 118

framework, vii, ix–x, 12, 19, 37, 48, 50, 54, 88, 94, 116, 135, 144, 147, 153–54, 156
freedom, 30
Freedom and Culture, 60
function, 16–17, 34, 41, 72, 77–78, 80–81, 84, 88, 96, 101, 106, 108, 111, 117, 126, 138
    functional system of value. *See* value, functional system of
future, 7, 33, 77–78, 84, 86, 92, 95, 102–3, 105, 113–14, 117, 121, 124, 127, 129, 131, 134–41, 143, 145
    generations, 33, 124, 129, 135–36, 138–39

global warming, 127, 129, 135
goal, 5, 11, 33, 77, 87, 92, 97, 103–4, 115, 131, 137
good(s), 4–5, 8, 13, 20, 25–26, 34–35, 38–41, 48–49, 53, 61–62, 68–70, 75–77, 80, 82, 84–88, 92, 94–95, 97, 101–2, 104, 106, 108, 110, 111, 114, 115, 117, 119, 120, 125–26, 131, 134–35, 144, 146, 156
    highest, 33, 46, 100, 115
    instrumental, 101, 138
    moral, 85, 102, 117, 121, 133
    natural, 70, 82, 85, 102, 107, 121, 133–34
    of a kind, 40
    of their own, 4, 8
    overall, 115, 119, 127, 135
good, the, 13, 30, 35, 39–40, 45–46, 48–49, 63, 75, 95, 110–11, 119–20, 124, 131, 133, 146, 156
greenhouse effect, 135
ground, grounding, viii, 2, 8–9, 14–15, 18, 20–21, 25, 34–35, 37, 42, 50–51, 63–65, 81, 94, 100, 104, 115, 124, 127, 137–38, 140, 144, 147, 151, 153, 155–56
    and consequent, 34
    foundational, 118
growth, 68–69, 77, 80, 84–86, 88–89, 94, 101, 111–12, 115–18, 120, 125, 127–28, 132, 135, 139, 152–53

habit(s), 71, 81, 84, 88, 98, 106, 110–12, 116, 130
　formation of, 71
habitats, 1–2, 9, 11–13, 15, 20, 25, 28, 43, 47, 62, 124, 126, 129, 132, 137–38, 143, 151
　unique, 138–39
happiness, 5, 101, 114, 129, 137–38
harmony, 29–30, 39, 77, 94, 116, 125, 136
health, 17, 27–31, 73, 82, 85, 87, 95, 101–2, 124, 126, 133, 138–39, 156
hedonism. *See* value, theories of
hierarchy, hierarchical, 17, 20, 25, 32, 42–44, 46–47, 49, 94, 104, 149
　hierarchy of values. *See* values, hierarchy of
historical, 8, 22, 24, 27, 40–41, 69, 74–75, 96, 149, 156
history, histories, 14–15, 40–41, 44, 69, 71, 75, 77, 82, 88, 135, 153, 156
　of philosophy, 14, 60
holism, vii–viii, 14–16, 25, 28–30, 49, 62, 77, 83, 126, 129, 140, 149–52, 156
　holistic model, 139–40
human, v–vii, 1–7, 10, 12–15, 17, 19–21, 23–24, 27–32, 34–38, 40–43, 45–52, 55–56, 57–65, 67–68, 71–73, 76–82, 85–89, 104, 106–8, 109–10, 114, 120, 122, 123, 126–38, 140, 143–44, 147–48, 150–57
　desires, 61
　nature, vii
　rights, 3, 24, 27, 32
Hume, David,, 20, 22–23, 27–28, 31, 35, 49, 56, 124
hypostatization, viii, 154
hypothesis, viii, 20, 151

idea, 4, 32, 45, 50–51, 60–62, 71, 73, 94–96, 99, 103–4, 112–13, 115, 138, 148, 150, 154, 157
ideal(s), xi, 23, 32, 39, 61, 76, 97, 102–3, 118, 145
idealism, viii, 91, 103–4, 156
　objective, 151

identity, 54, 92
imagination, 73, 113, 136
impact, 124, 126–29, 131–32, 135, 140–41. *See also* environmental impact
impulse, 71, 75, 78, 81, 84, 99, 110, 112, 116
indeterminate, 71
individuals, 1–2, 5–6, 9–10, 13, 15–17, 21–33, 35–36, 39–42, 44–47, 49–51, 53–55, 61–62, 109, 111, 124–25, 132, 134–36, 139–40, 152
induction, 85, 120
inherent value. *See* value, inherent
instance(s), 2, 21, 94
instincts, 71
institution(s), 59
instrumental, v, vi, 2–3, 5–6, 11–13, 17–18, 21, 24, 26, 28, 34, 36–37, 40–46, 49, 51, 57–59, 64, 84–85, 88, 95–96, 100–3, 105–8, 110, 112, 115, 117–18, 120–21, 125, 129–30, 134, 138, 140, 143–47, 152. *See also* values, instrumental
instrumentalism, 37, 59
integrity, 29
intellect, 71, 98
intellectualism, 68
intelligence, 27, 60, 81–82, 111, 113, 116, 119, 121–22, 123, 125–26, 132, 135–36
intent, 104
interaction, 12, 14, 19, 61, 72–75, 78–80, 82, 144, 149, 154–55
interest(s), 4, 7, 17, 27–28, 32, 38–39, 41, 47–48, 50, 54–55, 58, 63, 83, 86, 104–5, 107, 111, 119, 122
interrelations, 27–28, 30, 33, 39, 41, 77, 110
intrinsic, v–ix, 1–3, 5, 8–31, 33–47, 49–56, 57, 61–65, 76–77, 89, 91–101, 103, 105–8, 115, 117–22, 123–25, 131, 134, 137, 139–41, 143–48, 151–52, 155, 157. *See also* value, intrinsic

James, William, vi, viii, x, 63, 73, 149, 153, 156

judgment, 19–23, 82, 93, 107–8, 113
justice, 102
justification, vii, 3–5, 8–9, 13, 15, 31, 42, 45, 51–54, 57, 63–65, 99, 103, 106, 116–17, 119–20, 122, 134, 139, 145

Kant, Immanuel, 3, 6–9, 20, 31, 44, 46, 48, 104, 148
Katz, Eric, 57, 61–63, 67, 81–82, 89, 107, 121, 145, 148
"Kinship" theories, 4
Kitaro, Nishida, 149
Knowing and the Known, ix–x, 153–55
knowledge, 19, 39, 52, 54, 60, 91, 95, 103, 125, 144, 147–48, 150, 154, 157

Land ethic, 1, 12, 16, 22, 24, 26–29, 31–33, 35, 140
landscapes, 2, 11–12, 15, 28, 38, 85, 137
language, 20, 42, 60, 79, 87, 108, 133
laws, 151
Leopold, Aldo, 1, 16, 22, 24, 26–29, 35, 37, 65
Lewis, Clarence Irving, viii, 2, 18, 44, 51, 63, 105, 145–48, 150–51, 156
life, v, ix, xi, 2, 6–7, 11–12, 15–16, 26, 28–30, 32, 37, 39–46, 48–49, 53, 55, 59–60, 68, 70–76, 78, 80, 82–88, 94–95, 97, 101–2, 109–10, 115–19, 121, 12–35, 137–39, 143, 146–47, 151–53, 156
  as a whole, 44, 83, 116, 146–47, 156
  human, 37–38, 47, 60, 79–80, 87, 126, 137–38, 148, 156
  nonhuman, vii, 42, 47, 143–44, 156
likings, 98
linguistic, xi
Locke, John, 73, 153
locus of value. *See* value, locus of
logic, 37, 63, 99
logical, 15, 21, 34, 63, 94, 102, 150, 154

making, 8, 37, 43–44, 53, 60, 86, 96, 99
management ethic. *See* ethics, management
mark, 43, 47, 52, 58, 67, 69, 76–77, 87, 89, 99, 105, 112, 117, 133, 136–37

materialism, xi, 151
matter, 13, 37, 47, 54, 68, 73–74, 82, 89, 93, 96, 112, 116, 121, 136, 145, 151
meaning, 23, 79, 87, 93–95, 97, 105, 109, 116, 117, 147, 153
means, 2–3, 6–7, 10–11, 15–20, 39, 41–42, 46, 60–61, 63–65, 69, 74–76, 78–79, 81, 92–108, 109, 111–19, 121, 125, 127, 131, 133–36, 140, 149–50
  means-end relation. *See* relation, means-end
mechanism(s), 83, 107
memory, 95
mental, 71–73, 80–82, 93
  processes, 72
metaphysics, metaphysical, viii–xi, 2, 5–6, 9, 12–13, 18–19, 21, 25, 33, 36, 39, 64–65, 67–68, 72–73, 77, 81–83, 89, 93, 119–20, 124, 145, 153, 155–57
  dualistic, 72
  naturalistic, 67, 73
  of experience, x, 73, 83, 153
  of the subject, viii, 72, 155–56
  process, 153
Metaphysics of Aristotle, 68
method(s), vii–ix, 60–61, 67, 74, 110, 112, 122, 124
  experimental, 113, 115, 121
  m. of intelligence, 122
Mill, John Stuart, 105, 117
mind, 13, 15, 18, 36, 43, 73–75, 79, 81–82, 86, 89, 94, 98, 104, 110, 127, 129, 140, 145, 150–51
  "emergent" theory of m., 74
model, vii–ix, 14, 16, 23, 30, 32, 36–37, 42, 48–49, 54, 57, 59–60, 62, 64, 68, 77, 79–80, 91, 97, 99–103, 107–8, 117, 119–22, 123–24, 137, 139–40, 147, 150–53, 155–57
  holistic, vii, 139–40
monism, 33
Moore, George, E., 5, 20, 31, 45, 63, 147
moral(s), vi, 6, 9, 28, 33–34, 56, 68, 75, 83, 97, 103–4, 107, 109–10,

moral(s) *(continued)*
    116–18, 121, 124–25, 128–31, 136, 139, 145, 155–56
  m. communities, 29, 32
  m. considerability, consideration, v–vi, 3–4, 7–12, 14–17, 20, 25–33, 35–36, 39, 45–46, 48, 50, 86–87, 111, 121–22, 124–25, 128–29, 130–33, 137, 139–40, 143, 145, 147, 152, 156–57
  m. deliberation, 115, 117, 119, 121
  m. goods, 102
  m. obligation, 4, 10, 24, 26, 35–36, 47, 56
  m. pluralism, 33
  m. responsibility, 3
moral sentiments (theory), 1, 20, 22–23, 25, 28, 32–35, 54
morality, "reflective," 61, 110

Native Americans, 13
natural goods. *See* goods, natural
natural process. *See* process, natural
naturalism, vi–vii, xi, 19, 22, 50, 57–59, 67, 139, 140, 147
  empirical n., 67
naturalistic fallacy, 31
naturalistic metaphysics. *See* metaphysics, naturalistic
nature, vi–viii, 2–3, 5, 8, 12–16, 18–20, 22, 24–25, 29, 31–46, 49–52, 55, 57–65, 67–77, 79–83, 85–89, 91, 94, 104, 107–8, 109, 120–22, 123–24, 129–31, 133, 135, 137, 139–41, 143–44, 146–48, 150–52, 155–56
  biological, 72
  human, vii, 29, 49, 50, 59, 61, 67–68, 71–72, 77–81, 88–89, 114, 120, 122, 129, 132, 147
  trait of, 76–77, 133, 137
needs, 41, 71, 86–87, 111, 130–31
niches, 14, 25, 27, 29, 39–44, 127
Nietzsche, Friedrich, 91
nonhuman, vii–viii, 1, 3–5, 8–9, 12–15, 17–23, 26, 29–30, 32, 34–35, 38–39, 41–42, 46–47, 51–52, 55, 58–59, 63–64, 68, 87–88, 124, 128, 132–33, 134, 141, 143–44, 147, 151–53, 156–57
norm(s), 6–7, 29, 31, 35, 39, 129, 131, 133–34, 139, 146, 153
  cultural, 49
normative, 7, 24, 31, 41, 45, 145
  sciences. *See* sciences, normative
Norton, Bryan, vii, 1, 50–55, 134, 138
novelty, novelties, 70, 75–76, 97

object, 2, 5–6, 12–13, 18–19, 21–23, 34–36, 39, 41, 46, 49, 51, 53, 55, 63, 65, 70, 73, 82–83, 85, 89, 91–93, 107–8, 111–12, 114, 119–20, 141, 144–47, 149–52
objective, viii, 2, 6–8, 15, 18–19, 28, 36, 38–42, 44–45, 49, 52–55, 57, 68, 74–75, 84, 87, 89, 98, 102, 107–8, 110, 112–14, 141, 144–48, 150–51
objectivity, 18–19, 37, 47, 68, 84, 141
obligation, obligatory, viii, 4, 8, 10, 15, 32–33, 35–36, 47–48, 51, 56, 63, 89, 116–18, 123, 125–26, 129–30, 132, 134–35, 137–40, 145, 147, 151–52
  moral, 4, 10, 24, 26, 32, 35–36, 47–48, 56, 124, 130, 132, 139–40, 143, 151
  to the environment, 124, 126–27, 129, 132, 139, 148
  to species, 129, 134
old growth forest, 127
ontology, ontological, ix, 5–6, 9, 16–17, 20–21, 23–24, 26, 34, 36, 42, 55, 83, 96, 102, 144–45, 149, 157
organic, 29, 42–43, 45–46, 64, 68–69, 71–73, 78–80, 82–84, 86–88, 101, 106, 110, 119, 121–22, 130–31, 141, 157
organism, 40, 42, 48, 67–68, 71–82, 84, 86–88, 101–2, 108, 115–19, 121–22, 128–31, 140, 154–55
  in an environment, 68, 79, 82
organization, 86–87, 106, 116
origin, 51, 54, 68–69, 79, 87, 103, 131, 148

INDEX

ought(s), 15, 20, 31, 37–39, 41, 47, 54, 76, 100, 104–5, 121, 127–28, 135, 138, 145, 148, 157
overgrazing, 128, 132

pan-psychism, 63, 151 (of Peirce)
paradigm, 13–14, 25, 59, 126
part, ix–x, 12–15, 20, 22, 24, 30, 32, 37, 39, 45–46, 52, 59–61, 64, 68–71, 73–77, 79–80, 82, 87–89, 91–92, 95–97, 100–1, 106–8, 113, 117–18, 121, 124, 126, 130–32, 136, 138–39, 141, 150, 152, 154
past, 13, 23, 60–61, 84, 86, 102, 105, 107, 110, 113, 125–26
Peirce, Charles S., viii, 63, 145, 148, 150–51
perception, 73–74, 144–45, 147, 151
Perry, Ralph B., 104, 111
perspective, x–xi, 16, 29, 38, 43, 46, 60, 92–93, 116
phenomena, 84, 104
philosopher(s), x–xi, 58
 environmental, 65, 123
philosophic, 64
philosophy, philosophies, v–xi, 1–3, 8, 12–15, 29, 33, 55, 58–61, 63, 67, 79, 82–83, 88–89, 91, 93, 98, 108, 109, 121, 123, 133, 138, 140–41, 143–44, 149, 153, 156–57
 environmental, vi
 first, 156–57
 Greek, 120, 124
 political, 59, 61
physics, 14, 19, 48, 129
plant(s), 41, 80, 87, 128, 137
Plato(nic), 20, 94, 96, 103, 149, 155
pleasure, 4, 19, 34, 54–55, 104–5, 146, 148, 151
pluralism, 33, 62, 83, 88, 133–34
 moral, 33
plurality, 82, 97, 105, 137
population, 126–27, 135, 143
 growth, 136, 143
possible, viii, 5, 22, 28, 43, 53, 70, 76, 77, 94, 102–3, 105, 111, 126, 128, 131, 133–34, 147

potential, 7, 34, 43–44, 106, 118, 131, 151, 157
potentialities, 19, 70, 71, 87, 118
power, 106, 146
practical, vii, xi, 65, 95, 99, 103–6, 109, 129, 153, 157
practice, xi, 7, 27, 29–30, 34–35, 49, 89, 98, 105, 109, 115, 121, 157
pragmatic, viii, x–xi, 10, 62, 64, 145, 148, 152–53, 155, 157
Prall, David, 104, 107, 111, 118
precarious, the, 69, 70, 76, 77, 95
predators, 10, 30, 39, 40, 49, 133, 140
prediction(s), 141
premise, vi, 8–9, 11, 26, 29, 33, 35, 41, 56, 58, 151
present, vii–viii, 5, 7, 17, 20, 38, 40, 61, 70, 77, 81, 84, 94, 96–97, 102–3, 109, 111–14, 117, 119, 126–27, 135–37, 147
preservation of species. *See* species preservation
prey, 14, 24, 38–39, 133–34, 140, 151
principle, 10, 33–35, 48–49, 52, 115, 131–34, 139, 157
 of selection, 10, 34–35
prizing, 70, 83, 95, 99, 113
problematic, 16, 19, 33, 35, 45, 61, 75, 78, 84, 92, 94, 97–98, 104–6, 110–21, 125–29, 131–32, 135–36, 138–40, 153, 155
 p. situation, 78, 84
process(es), x, 7, 17, 24, 37–38, 40–41, 43, 45, 49–50, 60–61, 68–80, 82–84, 86, 87–89, 93–97, 101, 104–7, 112–17, 120, 122, 124–25, 129–30, 132, 135–36, 140–41, 143, 153–55
 mental, 72
 natural, vii, 24, 38, 43, 49, 50, 68, 70, 72, 75–77, 80, 82, 88–89, 106–8, 121–22, 133, 135, 139, 141
project(s), 8, 14, 37, 39, 40, 46, 57, 61, 68, 72, 76, 81, 88–89, 109, 124, 132–33, 150, 155
property, 6, 7, 19, 42, 79, 98, 106–7, 138, 144, 146

psychology, psychological, 31, 49, 81–84, 89, 104, 107, 122
   p. states, 7, 20, 34
purpose, 65, 106

quality, 19, 37, 39, 41–42, 45, 49, 69–70, 75–77, 83, 88–89, 95–96, 98, 105, 145–47, 152, 157

Radical Empiricism, 149
rationalism, 20, 105
rationalist(s), 8
realist(s), 44, 145
reality, 60, 68, 74, 101, 148, 156
realm, 1, 8, 12, 73–75, 77, 85, 103, 120, 141
reason, v, 8, 19–20, 22–23, 30–31, 51, 60–61, 71, 74, 106, 113, 125, 140, 143
reference, 5–6, 9, 19, 21, 42, 49, 52, 64–65, 67, 76, 95, 119, 125, 147, 149–50, 154
reflexive aspect (of relation of means to ends), 11, 17
Regan, Tom, vi, 1, 3–12, 14–18, 23–24, 26–27, 32, 35–37, 39, 41–42, 44, 46, 48, 50, 53, 55–56, 156
relation(s), relational, viii, 2, 6, 8, 10–12, 15, 17–26, 31, 34, 36, 39, 41–42, 46, 51, 53–56, 59–60, 63–65, 67, 69–70, 72–73, 77, 79, 82–83, 85, 91–93, 96–103, 106–8, 112–17, 119–22, 123–25, 134, 140–41, 145–48, 150, 152, 156–57
   aspect, 11–12, 34
   extrinsic-intrinsic value relation, 92
   instrumental, 34, 43, 64, 121, 140
   internal, 25
   means-end, 17, 91, 100, 122, 140
   social, 33, 37, 118
   systematic, 122, 156
relativism, relativist, 31, 35, 57, 62–63, 107, 145
religion, 149
result, 4, 51–52, 55, 76, 84, 93, 97, 99, 101–3, 115–16, 121, 126–27, 132, 135, 145
revaluation, 99, 133

right(s), v–vi, 5, 10, 13, 29, 35, 61, 104, 116, 119, 130, 132, 138–39, 148
   bearer of, v
   human, 3, 24, 27, 32
Rolston, Holmes, III, vi–vii, 1, 9–11, 18, 26, 36–50, 53, 55–56, 59, 64–65, 121, 123–24, 133, 135, 140, 152
root, 12, 22, 29, 40, 60, 78, 80, 89, 107, 120, 124, 143
Rorty, Richard, x–xi, 28
rule(s), 33, 54, 134

satisfaction(s), 51, 70, 81, 86–87, 95, 100, 114, 116–17, 138, 145, 148, 153
Schiller, F. C. S., viii–ix, 63, 151, 153–54, 156–57
science(s), 14, 15, 19–20, 23, 29–30, 32–33, 39, 48–49, 57, 68, 74, 86, 110, 121, 145, 154
   impact (on the environment). *See* impact
   normative, 145
   s. of ecology, 14, 23, 29–30
self-evident, 4, 95
self-development, 116, 118
self-realization, 26, 116, 118
senses, 75
sentient, 4, 53, 129, 152
Shaner, David E., 149
Singer, Peter, 3–4, 53, 124
situation, 68, 72, 78, 84–85, 92–94, 96–97, 99–106, 110–16, 118–22, 125–27, 132, 136–37, 150, 155
   problematic, 61, 78, 84, 92, 94, 97, 104–6, 110–21, 125–28, 131–32, 135–36, 138–39
social, viii, 22–23, 33, 37, 57–61, 64, 68, 71–72, 74–75, 78–80, 84–85, 87–88, 94, 107, 109–11, 116, 118, 121, 128, 132–34, 138–40, 150
   communication, 72
   conditions, 111, 121
   consequences, 116, 118
   environment, 74, 79–80, 87
   relations, 37
society, 27, 31, 46, 59, 68, 79, 116, 118, 128, 132, 135, 138, 140

soil. *See* air, soil and water
species, v–vi, 1–5, 8–11, 13–18, 20, 22–28, 30–31, 33, 35–40, 42–44, 47, 49–55, 58, 62, 70–72, 80, 85–88, 102, 120, 124, 126–27, 130–37, 139–40, 151
   climax, 151
   s. difference, 2
   endangered, 15, 31, 36, 48, 143, 148
   essential, 132, 134
   pioneer, 151
   s. preservation, 3, 10, 11, 15, 25, 27–28, 31, 33, 47, 50, 52, 53, 54, 56, 62, 133–35
   threatened, 132
speciesist, vi, 3, 20, 34
spirit, vii, xi, 133
stable, stabilities, 69–70, 77
standard(s), 25, 36, 41–42, 54, 84, 95, 103, 105, 113, 119, 132, 144, 151, 153
state, 2–4, 6, 54, 82, 96, 104, 115, 120, 145
Stevenson, Charles, 91
structure(s), v, 41, 71, 79, 152
subject(ive), vii–viii, xi, 2–4, 6–9, 13, 15, 17–23, 25–26, 34, 36–37, 39–45, 48–51, 54–55, 57–59, 62–64, 67–68, 72–73, 79, 81–83, 85–87, 89, 96–98, 103–5, 107–8, 111, 117, 119–22, 123–24, 126, 130, 140–41, 143–51, 153–57
   conscious, 12, 18, 49, 150
   subjective-objective distinction. *See* distinction, subjective-objective
subjectivity, viii–ix, 6, 18, 21, 38, 41, 43–44, 47–48, 51, 81, 107–8, 144–45, 147–50, 155
subordination, 103, 137, 151, 155
summum bonum, 100, 102, 117, 119, 139, 146–47. *See also* highest good
survival, vi, 29, 52, 60, 72, 83, 86–88, 122, 131–32, 134–36, 143
system(s), x, 16, 25, 30, 40–47, 52, 63–64, 67, 69, 74, 82, 104, 119, 134
systematic(s), x, 12, 122, 156
   relations, 122
systematically, 89

Taylor, Bob, 57, 67
Taylor, Paul, 53, 57–59, 72, 80, 82, 89
technology, 37, 78, 80, 126
teleological, 11, 17, 42, 53, 92
teleological aspect/dimension of intrinsic value, 11, 17
temporal, 24, 92, 96, 100, 135
theoretical, 120, 124, 148, 156
theory, -(ries), vi–ix, xi, 1–4, 6, 8–11, 13, 15–20, 22–29, 33, 35–38, 44–46, 48–56, 58–59, 61–63, 65, 69, 74–75, 79–80, 82–83, 85–86, 88–89, 91–94, 96–99, 102–8, 110–12, 114–15, 117, 120, 122, 124, 126, 137, 140–41, 143, 145–48, 151–52, 155, 157
   t. and practice, 102, 124
      separation of t. and p., 109, 120, 124
   environmental t.
      biocentric, 2
      ecocentric, 2
   ethical, 25, 33
   of value. *See* values, theory of
therapy, therapeutic, x–xi
thinking, 6, 12, 25, 35, 46, 59, 71, 87, 97, 148
thought, x, 9, 14, 32, 48, 58–59, 71, 93, 95, 99, 103, 109, 121, 131, 140, 151
tradition, 5, 8, 23, 30, 43, 46, 50, 53–54, 57–58, 60–61, 73, 77, 83, 89, 92–93, 95, 98, 102–3, 105–6, 109, 115–16, 119, 121, 144, 150, 153, 155, 157
transaction, transactionalism, viii–x, 154
transformative value. *See* value, theory of, transformative
truth, 20, 50, 150
Tufts, James, ix

unchanging, 97
unconscious, 60
universal(s), 6, 27, 31, 34, 68, 87, 121
universality, 31, 148
universe, 88, 108, 151
   pluralistic, 68, 150

utilitarian(ism), 3, 13, 26, 35, 37, 53–54, 104, 117, 148
utility, 5, 13, 17, 21, 37, 55, 62

valuable in itself, valuable for its own sake, 21, 145. *See also* for its own sake
valuation, 6, 63, 65, 78, 80, 85, 92, 98, 110–14, 120, 156
value(s), v–viii, 1–3, 5–30, 32–33, 35–56, 57, 61–65, 68, 70, 75–78, 80–89, 91–92, 94–108, 109, 111–16, 118–22, 123–26, 128–41, 143–48, 150–53, 155–57
  anthropocentric, v, 52, 62
  bearers of, 3, 55
  degrees of, 6, 43–44
  hierarchy of, 43, 46, 94
  extrinsic, 92, 99, 145–47
  inherent, 2, 5–7, 9–11, 18, 22–23, 99, 145–48, 151
  instrumental, vi, 3, 5, 11–13, 21, 24, 26, 28, 34, 36, 40, 42, 44–45, 49, 51, 84–85, 88, 96, 100–1, 103, 106–8, 120, 125, 129–30, 134, 144–47, 152
  intrinsic, v–ix, 1–3, 5, 8–31, 33–47, 49–56, 57, 61–65, 77, 89, 91–101, 105–8, 115, 117–22, 123–25, 131, 134, 137, 139–40, 143–48, 151–52, 155, 157
  esthetic theories of, 9
  locus of, 1–2, 5, 9–10, 13, 15, 17, 21, 23–24, 35, 39, 44–45, 50, 53, 55–56
  natural, 43
  of the whole, 15, 26, 30, 40, 42, 44, 124, 152
  scope of, 2, 36, 82, 85, 87–88, 147, 155–56
  wild, 38–39, 45, 47
value, Dewey's theory of
  includes bringing about (ends/values), 70, 75, 82–83, 95, 106–7, 127
  Instrumental, 91, 134

  as positive direction of change, 84, 94, 117–19, 125–27, 131, 135–37, 141
  See also growth; values, instrumental
value-free, 19, 20, 23, 32, 39, 48
value judgments, 32, 91
value theory, value(s), theory/theories of, vi–viii, 1–2, 5, 8, 12–13, 17–19, 26, 28, 36, 38, 41, 44, 48, 50–51, 54–56, 57–59, 61–63, 75, 83, 81, 89, 92–93, 98, 104, 105–8, 111–12, 114, 122, 133, 141, 143, 145, 151, 155
  conativism, 27
  consequentialist, vii, 56, 104
  Demand value, 51
  functional system of, 16
  hedonism, 104, 146
  Interest theory, 104
  esthetic, 9
  non-anthropocentric, v
  noncognitivist, 91
  relational, viii, 2, 19, 34, 39, 51, 55, 85, 107, 108, 120, 146, 148, 145–46, 148
  Transformative, 52
valuer, 6, 22–23, 34, 51, 63, 148
virtue, 6, 99, 103, 120, 131

water. *See* air, soil, water
web of relations, 14, 44–45, 47, 59, 64–65, 83, 85, 119–21, 124–25, 127–28, 132–33, 140, 152
Weston, Anthony, 57–58, 63–65, 89, 145
wetlands, 137–38
whole, v–ix, xi, 6, 8, 13–16, 19–20, 22, 24–33, 35–36, 38–50, 55, 58–59, 61, 64, 67–68, 76, 78, 81–82, 84, 86, 88, 95–96, 101, 105, 108, 112, 114–15, 117, 119–22, 124–39, 143, 146–52, 156
wild, 11, 13, 16, 25, 27, 30–33, 36, 38–40, 43–49, 52–53, 55, 58–59, 65, 126, 147
wildlife, 13, 137

world, viii, xi, 12–15, 19, 22, 30, 37, 39, 47–49, 52, 55, 62, 64, 68, 70, 73–77, 79–80, 86, 103, 107–8, 109, 118, 120, 124, 128–29, 131, 137, 139, 141, 144–45, 149, 151–53, 156–57
  w. view, 13, 15, 19, 33, 48–49, 52, 149, 152
    ecocentric, 149
worth, 21, 23, 53, 95, 99, 105, 108, 114, 136

zones, 43

www.ingramcontent.com/pod-product-compliance
Lightning Source LLC
Chambersburg PA
CBHW020649230426
43665CB00008B/368